Lecture Notes in Mathematics

1220

Séminaire d'Algèbre Paul Dubreil et Marie-Paule Malliavin

Proceedings, Paris 1985
(37ème Année)

Edité by M.-P. Malliavin

Springer-Verlag
Berlin Heidelberg New York London Paris Tokyo

Editeur

Marie-Paule Malliavin
Université Pierre et Marie Curie, Mathématiques
10, rue Saint Louis en l'Ile, 75004 Paris, France

Mathematics Subject Classification (1980): 03G20, 13H10, 16A08, 16A15, 16A38, 16A42, 16A46, 16A48, 16A64, 17B65, 18G15

ISBN 3-540-17185-1 Springer-Verlag Berlin Heidelberg New York
ISBN 0-387-17185-1 Springer-Verlag New York Berlin Heidelberg

Printing and binding: Druckhaus Beltz, Hemsbach/Bergstr.
2146/3140-543210

Liste des auteurs.

*

TABLE DES TITRES.

* *

PREVIOUS VOLUMES OF THE "SEMINAIRE PAUL DUBREIL" WERE PUBLISHED IN THE LECTURE NOTES , VOLUMES 586 (1976), 641 (1977), 740 (1978), 795 (1979), 867 (1980), 924 (1981), 1029 (1982) and 1146 (1983-84).

DYNKIN ALGEBRAS

Dieter Happel

Let k be an algebraically closed field. We will consider finite dimensional basic k-algebras A and we want to describe the derived category of bounded complexes over mod A, the category of finitely generated right A-modules. In the general situation this is a rather ambitious goal and here we restrict to a rather special class of algebras, namely the class of Dynkin algebras. For other classes of algebras we refer to [9], [11].

For the definition of Dynkin algebras we have to recall certain notions.

A cycle in mod A is given by a sequence $X_o \xrightarrow{f_o} X_1 \xrightarrow{f_1} X_2 \longrightarrow \dots \xrightarrow{f_{n-1}} X_n = X_o$ of indecomposable A-modules X_i and non-zero maps f_i which are not isomorphisms. In case there are no cycles in mod A we say that mod A is directed. It is shown in [14] that a finite dimensional k-algebra A with directed module category is necessarily representation-finite, i. e. A admits up to isomorphism only a finite number of indecomposable A-modules.

We recall the definition of the Cartan matrix associated to A. Denote by $P_1,\dots,P_n$ a complete set of isomorphism classes of indecomposable projective A-modules. The Cartan matrix C_A is an integer-valued $(n \times n)$-matrix with entries

$$c_{ij} = \dim_k \operatorname{Hom}_A(P_i,P_j) \qquad 1 \leq i, j \leq n .$$

We say that two integer-valued square matrices M,N of the same size are congruent $(M \sim N)$ if there exists an integer-valued matrix g, which is invertible over the integers such that $M = gNg^t$, where g^t denotes the transpose of g.

We are interested in a special congruence class. For this let Δ be a Dynkin diagram of type $\mathbb{A}_n$, $\mathbb{D}_n$, $\mathbb{E}_6$, $\mathbb{E}_7$, $\mathbb{E}_8$. In this article the notion Dynkin diagram always refers to one of these diagrams. Let $\vec{\Delta}$ be a quiver with underlying graph Δ and let $k\vec{\Delta}$ be the path algebra of $\vec{\Delta}$ (see for example [5]). Then using reflections it is easily seen that $C_{k\vec{\Delta}} \sim C_{k\vec{\Delta}'}$, if and only if the underlying graphs of $\vec{\Delta}$ and $\vec{\Delta}'$ coincide. So we may just use the notation C_Δ as an abbreviation for the set of matrices $C_{k\vec{\Delta}}$ for the different orientations of Δ.

Now we are ready to give the definition of a Dynkin algebra.

A finite dimensional basic k-algebra A is called a Dynkin algebra provided mod A is directed and $C_A \sim C_\Delta$ for some Dynkin diagram Δ.

We call Δ also the type of the Dynkin algebra A.

The aim of this article is to outline a characterization of Dynkin algebras in terms of derived categories. To be more precise denote by $D^b(\text{mod } A)$ the derived category of bounded complexes over mod A [15].

Theorem: Let A be a finite dimensional k-algebra and let Δ be a Dynkin diagram. Then the following are equivalent.

(i) $D^b(\text{mod } A) \approx D^b(\text{mod } k\vec{\Delta})$ as triangulated categories

(ii) A is a Dynkin algebra of type Δ.

This article gives an outline of the necessary results needed in the above characterization. For a more detailed account we refer to [9].
In section 1 we study the derived category for module categories and determine $D^b(\text{mod } k\vec{\Delta})$ for some quiver $\vec{\Delta}$. Section 2 gives a representation-theoretic version of the root system corresponding to the Dynkin diagram Δ. And we quote from [7] that Dynkin algebras are iterated tilted algebras. Section 3 gives the proof of the above mentioned characterization and in section 4 we conclude by giving some examples.

For the results on triangulated categories and derived categories which are needed here we refer to [15].
Throughout the composition of maps $f : X \to Y$, $g : Y \to Z$ is denoted by fg.
The occuring quivers are assumed to be connected.

1. Derived categories of module categories

In this section we want to extend certain useful notions in the representation theory of finite-dimensional algebras to the derived category.

1.1. Dimension vectors and a bilinear form

We keep the notation from the introduction. Denote by $C^b(\text{mod } A)$ the category of bounded complexes over mod A. To an object $X^\cdot = (X^n, d^n)$ in $C^b(\text{mod } A)$ we associate a dimension vector by $\underline{\dim}\, X^\cdot = \sum_{i \in \mathbb{Z}} (-1)^i \underline{\dim}\, X^i$, where $\underline{\dim}\, X^i$ is the usual dimension vector for X^i considered as A-module. This is by definition the vector whose j-th component is given by $\dim_k \operatorname{Hom}_A(P_j, X^i)$, where $P_1, \ldots, P_n$ are a complete set of representatives of the isomorphism classes of indecomposable projective A-modules.

We want to extend this definition to the derived category $D^b(\text{mod } A)$. For this we have the following lemma.

Lemma: Let $X^\cdot, Y^\cdot \in D^b(\text{mod } A)$ and $X^\cdot$ isomorphic to $Y^\cdot$. Then $\underline{\dim}\, X^\cdot = \underline{\dim}\, Y^\cdot$.

Proof: Let $u^\cdot : X^\cdot \to Y^\cdot$ be a quasi-isomorphism. It is clearly enough to show that in this case $\underline{\dim}\, X^\cdot = \underline{\dim}\, Y^\cdot$. In fact consider the triangle $X^\cdot \xrightarrow{u^\cdot} Y^\cdot \to Z^\cdot \to TX^\cdot$.
This yields the long exact sequence for the cohomology groups

$$\ldots\ H^i(X^\cdot) \xrightarrow{H^i(u^\cdot)} H^i(Y^\cdot) \to H^i(Z^\cdot) \to H^{i+1}(X^\cdot) \xrightarrow{H^{i+1}(u^\cdot)} H^{i+1}(Y^\cdot) \quad \ldots$$

Since $u^\cdot$ is a quasi-isomorphism $H^i(u^\cdot)$ is an isomorphism for all $i \in \mathbb{Z}$. Thus we see that $H^i(Z^\cdot) = 0$ for all $i \in \mathbb{Z}$. In other words $\underline{\dim}\, Z^\cdot = 0 \in \mathbb{Z}^n$. Since $\underline{\dim}\, TX^\cdot = -\underline{\dim}\, X^\cdot$ as an immediate calculation shows we obtain:

$$\underline{\dim}\, Y^\cdot - \underline{\dim}\, X^\cdot = \underline{\dim}\, Y^\cdot + \underline{\dim}\, TX^\cdot = \underline{\dim}(Y^\cdot \oplus TX^\cdot) = \underline{\dim}\, Z^\cdot = 0 .$$

Thus $\underline{\dim}\, X^\cdot = \underline{\dim}\, Y^\cdot$.

The lemma shows that we can define a dimension vector for the objects in $D^b(\mathrm{mod}\, A)$.

For the rest of this section we assume that A has finite global dimension. In this case the Cartan matrix C_A is invertible and $\chi_A = C_A^{-1}$ is usually called the Euler-characteristic. This gives a bilinear form $\langle -,- \rangle$ on $\mathbb{Z}^n$ by $\langle x,y \rangle = x^t \chi_A y$.

This bilinear form has the following homological interpretation (for a more detailed account we refer to [14]). Let X,Y be A-modules. Then

(*) $$\langle \underline{\dim}\, X, \underline{\dim}\, Y \rangle = \sum_{i \geq 0} (-1)^i \dim_k \mathrm{Ext}_A^i(X,Y) .$$

The next lemma extends this to the derived category. For a proof we refer to [9].

Lemma: Let $X^\cdot, Y^\cdot \in D^b(\mathrm{mod}\, A)$. Then

(**) $$\langle \underline{\dim}\, X^\cdot, \underline{\dim}\, Y^\cdot \rangle = \sum_{i \in \mathbb{Z}} (-1)^i \dim_k \mathrm{Hom}_{D^b(\mathrm{mod}\, A)}(X^\cdot, T^i Y^\cdot) .$$

Recall that the two bilinear forms $\langle\ ,\ \rangle_1$, $\langle\ ,\ \rangle_2$ on $\mathbb{Z}^n$ and $\mathbb{Z}^m$ are called congruent if there exists a linear map $f : \mathbb{Z}^n \to \mathbb{Z}^m$ such that for all $x,y \in \mathbb{Z}^n$ we have $\langle x,y \rangle_1 = \langle f(x), f(y) \rangle_2$.

Proposition: Let A,B be finite dimensional k-algebras of finite global dimension and suppose that $D^b(\mathrm{mod}\, A) \xrightarrow{F} D^b(\mathrm{mod}\, B)$ is an equivalence of triangulated categories. then there exists $f \in GL_n(\mathbb{Z})$ such that for $x,y \in \mathbb{Z}^n$ we have $\langle x,y \rangle_A = \langle f(x), f(y) \rangle_B$ and the number of different simple A-modules equals the number of different simple B-modules. Moreover, for $X^\cdot \in D^b(\mathrm{mod}\, A)$ we have $\underline{\dim}\, F(X^\cdot) = f(\underline{\dim}\, X^\cdot)$.

Proof: Again denote by $P_1,\ldots,P_n$ the indecomposable projective A-modules and let $M_i^\cdot = F(P_i)$ for $1 \leq i \leq n$. We claim that $\langle \underline{\dim}\, P_i, \underline{\dim}\, P_j \rangle_A = \langle \underline{\dim}\, M_i^\cdot, \underline{\dim}\, M_j^\cdot \rangle_B$. In fact: $\langle \underline{\dim}\, P_i, \underline{\dim}\, P_j \rangle_A = \dim_k \mathrm{Hom}_A(P_i,P_j)$ and

$$\begin{aligned} \langle \dim M_i^\cdot, \dim M_j^\cdot \rangle_B &= \sum_{r \in \mathbb{Z}} (-1)^r \dim \mathrm{Hom}_{D^b(\mathrm{mod}\, B)}(M_i^\cdot, T^r M_j^\cdot) \\ &= \sum_{r \in \mathbb{Z}} (-1)^r \dim \mathrm{Hom}_{D^b(\mathrm{mod}\, A)}(P_i, T^r P_j) \\ &= \dim_k \mathrm{Hom}_A(P_i,P_j) . \end{aligned}$$

We obtain a map $f : \mathbb{Z}^n \to \mathbb{Z}^m$ by defining for $x = \sum_{i=1}^n \mu_i \underline{\dim}\, P_i$ the value for

$f(x) = \sum_{i=1}^{n} \mu_i \underline{\dim} M_i^{\cdot}$, where $\mathbb{Z}^m = K_o(B)$ and m is the number of different simple B-modules. Obviously f is a congruence.
Moreover, f is surjective. This holds since all simple B-modules are in the image of F. In particular $n \geq m$.
Using the inverse functor to F we may show in the same way that $m \geq n$.

Thus the number of different simple A-modules equals the number of different simple B-modules.

To show the equality $\underline{\dim} F(X^{\cdot}) = f(\underline{\dim} X^{\cdot})$ for $X^{\cdot} \in D^b(\text{mod } A)$ it is clearly enough to show it for $P^{\cdot} \in K^b(\mathcal{P}_A)$, where $\mathcal{P}_A$ denotes the full subcategory of mod A consisting of projective A-modules. For this we define the latitude of a complex. Let $X^{\cdot} = (X^n, d^n)$. There exist $i \leq j$ such that $X^r = 0$ for $r < i$ and $X^s = 0$ for $r > j$ and $X^i \neq 0 \neq X^j$. Then the latitude of $X^{\cdot}$ is $\ell(X^{\cdot}) = j-i+1$. We prove now the assertion by induction on the latitude of $P^{\cdot} \in K^b(\mathcal{P}_A)$. For $\ell(P^{\cdot}) = 1$ this is just the definition of f. Observe that F commutes with the translation functor. Now let $\ell(P^{\cdot}) = r$. We may assume that $P^{\cdot} = (P^n, d^n)$ satisfies $P^o \neq 0$ and $P^n = 0$ for $n < 0$. Let $P^{\cdot\prime}$ be the truncated complex, with $(P^{\cdot\prime})^o = 0$ and $(P^{\cdot\prime})^n = P^n$ for $n \neq 0$. Then we obtain a triangle $T^{-}P^o \to P^{\cdot\prime} \to P^{\cdot} \to P^o$. And we obtain:

$$\underline{\dim} F(P^{\cdot}) = \underline{\dim} F(P^{\cdot\prime}) + \underline{\dim} F(P^o) = f(\underline{\dim} P^{\cdot\prime}) + f(\underline{\dim} P^o) = f(\underline{\dim} P^{\cdot})$$

using induction and the fact that F is exact.
This finishes the proof of the proposition.

1.2. Tilting theory

Let us start by recalling the basic definitions. Following [10] a module M_A is called a <u>tilting module</u> provided the following conditions are satisfied:

(i) $\text{proj.dim } M_A \leq 1$
(ii) $\text{Ext}_A^1(M_A, M_A) = 0$
(iii) There exists a short exact sequence $0 \to A_A \to M' \to M'' \to 0$ with M', M'' being direct sums of summands of M.

To a tilting module M_A we associate the algebra $B = (\text{End } M_A)^{op}$.

Let A, B be finite dimensional algebras. We say that A and B are <u>tilting equivalent</u> provided there exists a family $(A_i, M_{A_i})_{0 \leq i \leq m}$ of finite dimensional k-algebras A_i and tilting modules M_{A_i} satisfying:

(i) $A_o = A$
(ii) $A_{i+1} = (\text{End } M_{A_i})^{op}$ $0 \leq i < m$
(iii) $A_m = B$.

There is a more restricted notion. Let A be an hereditary k-algebra. Then there exists a quiver $\vec{\Delta}$ such that A is given by the path algebra $k\vec{\Delta}$.
A finite-dimensional k-algebra B is called an iterated tilted algebra of type $\vec{\Delta}$ provided $k\vec{\Delta}$ and B are tilting equivalent and there exists a family $(A_i, M_{A_i})_{0 \le i \le m}$ as above satisfying the additional property:

(iv) An indecomposable A_{i+1}-module $X_{A_{i+1}}$ is of the form $\mathrm{Hom}_{A_i}(M_{A_i}, X_{A_i})$ or $\mathrm{Ext}^1_{A_i}(M_{A_i}, X_{A_i})$ for some indecomposable A_i-module X_{A_i}.

The tilting theory asserts that if A and B are tilting equivalent then mod A and mod B are closely related (compare [2], [10]). For example let us mention one result of Bongartz. If gl dim $A < \infty$ then also gl dim $B < \infty$. It is this experience which motivated the following result. For a detailed proof we refer to [8] or [9].

Theorem: Let A be a finite dimensional k-algebra and assume that gl dim $A < \infty$. Let B be a finite dimensional k-algebra with A and B tilting equivalent then $D^b(\mathrm{mod}\ A)$ is equivalent to $D^b(\mathrm{mod}\ B)$ as triangulated category.

Proof: It is enough to show the assertion for a tilting triple (A, M_A, B) where M_A is a tilting module and $B = (\mathrm{End}\ M_A)^{op}$. We will just give the functor which gives the equivalence and refer to [8] for details. Since gl dim $A < \infty$ we have that $D^b(\mathrm{mod}\ A) \approx K^b(I_A)$ where I_A denotes the full subcategory of mod A consisting of the injective A-modules. We first define a functor F from $K^b(I_A)$ to $K^b(\mathrm{mod}\ B)$. For $I^\cdot = (I^n, d^n) \in K^b(I_A)$ let $F(I^\cdot)$ be the complex $((\mathrm{Hom}_A(M_A, I^n))^n, (\mathrm{Hom}_A(M_A, d^n)^n)$. Denote by Q_B the localization functor from $K^b(\mathrm{mod}\ B)$ to $D^b(\mathrm{mod}\ B)$ then $\tilde{F} = Q_B \cdot F$ is a functor from $K^b(I_A)$ to $D^b(\mathrm{mod}\ B)$.

Using this theorem and the proposition in 1.1 we obtain the following corollary which is already contained in [10].

Corollary: Let A, B be tilting equivalent finite dimensional k-algebras of finite global dimension. Then $\langle\ ,\ \rangle_A$ is congruent to $\langle\ ,\ \rangle_B$.

1.3. Auslander-Reiten triangles

In analogy to the theory of Auslander-Reiten sequences in module categories [1] we introduce the notion of an Auslander-Reiten triangle in $D^b(\mathrm{mod}\ A)$.
Let $X^\cdot, Z^\cdot$ be indecomposable objects in $D^b(\mathrm{mod}\ A)$. A triangle
$X^\cdot \xrightarrow{u^\cdot} Y^\cdot \xrightarrow{v^\cdot} Z^\cdot \xrightarrow{w} TX^\cdot$ is called an Auslander-Reiten triangle provided the following two conditions are satisfied:

(i) $w \neq 0$

(ii) For all indecomposable objects $V^\cdot \in D^b(\mathrm{mod}\ A)$ and morphisms $f^\cdot : V^\cdot \to Z^\cdot$

such that $f^\cdot$ is not an isomorphism there exists $g^\cdot : V^\cdot \to Y^\cdot$ with $f^\cdot = g^\cdot v^\cdot$.

Recall that a morphism $\varphi : X \to Y$ in an arbitrary additive category is called <u>irreducible</u> if φ is not an isomorphism but for any factorization $\varphi = \psi_1\psi_2$ we have that either ψ_1 is a coretraction or ψ_2 is a retraction.
Denote by ν the <u>Nakayama functor</u> from $K^b(P_A)$ to $K^b(I_A)$. Then we obtain a functor $\tau = T^{-1}\nu$ from $K^b(P_A)$ to $K^b(I_A)$. If gl dim $A < \infty$ then we may consider τ as endo-functor on $D^b(\text{mod } A)$. This is even a selfequivalence on $D^b(\text{mod } A)$.

<u>Theorem</u>: <u>Suppose</u> A <u>is a finite dimensional</u> k-<u>algebra of finite global dimension and let</u> $Z^\cdot$ <u>be indecomposable in</u> $D^b(\text{mod } A)$. <u>Then there exists an Auslander-Reiten triangle</u> $\tau Z^\cdot \xrightarrow{u^\cdot} Y^\cdot \xrightarrow{v^\cdot} Z^\cdot \xrightarrow{w^\cdot} T\tau Z^\cdot$.

<u>Proof</u>: This is analogous to the proof given in [5].

In the next proposition we collect some basic properties of Auslander-Reiten triangles. The proof is straightforward and we refer to [9] for details. But we want to stress that in the proof we were guided by the approach in [5].

<u>Proposition</u>: Suppose A is a finite dimensional k-algebra of finite global dimension and let $X^\cdot \xrightarrow{u^\cdot} Y^\cdot \xrightarrow{v^\cdot} Z^\cdot \xrightarrow{w^\cdot} TX^\cdot$ be an Auslander-Reiten triangle. Then

(i) Auslander-Reiten triangles are unique up to isomorphisms

(ii) For all indecomposable objects $W^\cdot \in D^b(\text{mod } A)$ and morphisms $g^\cdot : X^\cdot \to W^\cdot$ such that $g^\cdot$ is not an isomorphism there exists $h^\cdot : Y^\cdot \to W^\cdot$ with $g^\cdot = u^\cdot h^\cdot$.

(iii) $u^\cdot$ is irreducible

(iv) $v^\cdot$ is irreducible

(v) Let $u_1^\cdot : X^\cdot \to Y^{\cdot\prime}$ be irreducible then there exists $Y^{\cdot\prime\prime}$ and $u_2^\cdot : X^\cdot \to Y^{\cdot\prime\prime}$ such that $u^\cdot = (u_1^\cdot, u_2^\cdot)$

(vi) Let $v_1^\cdot : Y^{\cdot\prime} : Y^{\cdot\prime} \to Z^\cdot$ be irreducible then there exists $Y^{\cdot\prime\prime}$ such $v_2^\cdot : Y^{\cdot\prime\prime} \to X^\cdot$ such that $v^\cdot = (v_1^\cdot, v_2^\cdot)$.

Thus Auslander-Reiten triangles carry the information on irreducible morphisms. We define the <u>quiver</u> $\Gamma(D^b(\text{mod } A))$ of $D^b(\text{mod } A)$ to be the quiver having as vertices the isomorphism classes of indecomposable objects in $D^b(\text{mod } A)$ and there is an arrow from an isomorphism class $[X^\cdot]$ to an isomorphism class $[Y^\cdot]$ provided there exists an irreducible morphism $f^\cdot : X^\cdot \to Y^\cdot$.

1.4. <u>The quiver of $D^b(\text{mod } k\vec{\Delta})$</u>

First we determine the indecomposable objects in $D^b(\text{mod } k\vec{\Delta})$ for some quiver $\vec{\Delta}$ without oriented cycles.

<u>Lemma</u>: Let $X^\cdot$ be an indecomposable object in $D^b(\mathrm{mod}\, k\vec{\Delta})$. Then $X^\cdot$ is quasi-isomorphic to a stalk complex with indecomposable stalk.

<u>Proof</u>: We show that the indecomposable objects in $K^b(\mathcal{I}_A)$ are of the form $\ldots 0 \to I^j \xrightarrow{d^j} I^{j+1} \to 0 \ldots$ with d^j epi and $\ker d^j$ is an indecomposable $k\vec{\Delta}$-module.

Suppose we have given $I^\cdot$ indecomposable in $K^b(\mathcal{I}_A)$. Up to translation we may assume that $I^\cdot$ is of the following form:

$$\ldots 0 \to I^0 \xrightarrow{d^0} I^1 \xrightarrow{d^1} I^2 \quad \ldots \quad \text{with } I^0 \neq 0 .$$

We claim that d^0 is epi and $d^j = 0$ for $j \geq 1$.

In fact, consider a factorization of d^0, $I^0 \xrightarrow{g} X \xrightarrow{h} I^1$ for some $k\vec{\Delta}$-module X and g epi and h mono. Then X is an injective $k\vec{\Delta}$-module. Thus h is split mono. So there exists a $k\vec{\Delta}$-module C and $u \in \mathrm{Hom}(C, I^1)$ such that $X \oplus C \xrightarrow{\binom{h}{u}} I^1$ is an isomorphism.

Consider now the morphism of complexes

$$\begin{array}{ccccc}
I^0 & \xrightarrow{\quad d^0 \quad} & I^1 & \xrightarrow{\quad d^1 \quad} & I^2 \ \ldots \\
\Big\| & & \Big\uparrow {\scriptstyle \binom{h}{u}} & & \Big\| \\
I^0 \oplus 0 & \xrightarrow{\begin{pmatrix} g & 0 \\ 0 & 0 \end{pmatrix}} & X \oplus C & \xrightarrow{\begin{pmatrix} 0 & 0 \\ 0 & ud^1 \end{pmatrix}} & 0 \oplus I^2 \ \ldots
\end{array}$$

Observe that $hd^1 = 0$. This morphism is an isomorphism. Thus we obtain the assertion.

Thus we may use the following notation for the indecomposable objects in $D^b(\mathrm{mod}\, k\vec{\Delta})$. Let X be an indecomposable $k\vec{\Delta}$-module. Denote by $X[i]$ for $i \in \mathbb{Z}$ the indecomposable object $T^i X$ in $D^b(\mathrm{mod}\, k\vec{\Delta})$. Then the previous lemma shows that an indecomposable object $X^\cdot$ in $D^b(\mathrm{mod}\, k\vec{\Delta})$ is isomorphic to $X[i]$ for some indecomposable $k\vec{\Delta}$-module X and some $i \in \mathbb{Z}$.

For the determination of the arrows in $\Gamma(D^b(\mathrm{mod}\, k\vec{\Delta}))$ we use the Auslander-Reiten triangles. Let $X[i]$ be an indecomposable object in $D^b(\mathrm{mod}\, k\vec{\Delta})$. If X is not projective then $\tau(X[i]) = (\tau X)[i]$ where τX is the Auslander-Reiten translate of X as an A-module. Thus let $0 \to \tau X \to Y \to X \to 0$ be the Auslander-Reiten sequence. This yields an Auslander-Reiten triangle $(\tau X)[i] \to Y[i] \to X[i] \to T((\tau X)[i])$.

Now let $X = P$ be an indecomposable projective module. Let $S = \mathrm{top}\, P$. Then $\tau(X[i]) = T^-(I[i])$ where I is an indecomposable injective module with $\mathrm{soc}\, I = S$. Let w be the canonical map $P \to I$ vanishing on $\mathrm{rad}\, P$. Then an easy computation shows that we obtain as Auslander-Reiten triangle in this case:

$$T^-(I[i]) \to T^-(I/S[i]) \oplus (\mathrm{rad}\, P)[i] \to P[i] \to I[i] .$$

From these observations we can read off the structure of $\Gamma(D^b(\mathrm{mod}\, k\vec{\Delta}))$. Denote by

$\Gamma[i]$ a copy of the Auslander-Reiten quiver of $\mathrm{mod}\ k\vec{\Delta}$ for $i \in \mathbb{Z}$. For the structure of $\Gamma[i]$ we refer to [13], [14]. Recall that the indecomposable projective modules as well as the indecomposable injective modules are indexed by the vertices of $\vec{\Delta}$. For a vertex a in $\vec{\Delta}$ we denote by $P(a)$ and $I(a)$ the corresponding projective and injective module.

<u>Proposition:</u> $\Gamma(D^b(\mathrm{mod}\ k\vec{\Delta})) = \bigcup_{i\in\mathbb{Z}} \Gamma[i]$ with additional arrows from $I(a)[i]$ to $P(b)[i+1]$ whenever there is an arrow from a to b in $\vec{\Delta}$.

As an immediate corollary we obtain the following. For the definition of $\mathbb{Z}\Delta$ we refer to [12].

<u>Corollary</u>: Let Δ be a Dynkin diagram. Then $\Gamma(D^b(\mathrm{mod}\ k\vec{\Delta})) = \mathbb{Z}\Delta$.

Even more is true. Denote by $\mathrm{ind}\ D^b(\mathrm{mod}\ k\vec{\Delta})$ the full subcategory of $D^b(\mathrm{mod}\ k\vec{\Delta})$ with objects representatives from the isomorphism classes of indecomposable objects in $D^b(\mathrm{mod}\ k\vec{\Delta})$. Moreover denote by $k(\mathbb{Z}\Delta)$ the mesh category of $\mathbb{Z}\Delta$ [3], [12]. Then $\mathrm{ind}\ D^b(\mathrm{mod}\ k\vec{\Delta}) = k(\mathbb{Z}\Delta)$. For this we refer to [9].

2. <u>Cylinders</u>

In this section we reproduce some results of [7]. First we present a representation-theoretic version of the root system corresponding to the Dynkin diagram Δ.
The square of the translation functor T on $\mathrm{ind}\ D^b(\mathrm{mod}\ k\vec{\Delta})$ gives rise to a Galois covering $\mathrm{ind}\ D^b(\mathrm{mod}\ k\vec{\Delta}) \xrightarrow{\pi} (\mathrm{ind}\ D^b(\mathrm{mod}\ k\vec{\Delta})/T^2 = \mathcal{R}(\vec{\Delta})$. For the covering techniques we refer to [3] and [6].
$\mathcal{R}(\vec{\Delta})$ is called the <u>root category</u>. It clearly does not depend on the orientation of Δ and will therefore be simply denoted by $\mathcal{R}(\Delta)$. Clearly $\mathcal{R}(\Delta)$ inherits from $\mathrm{ind}\ D^b(\mathrm{mod}\ K\vec{\Delta})$ the structure of a stable translation quiver. In the following we will identify $\mathcal{R}(\Delta)$ with its quiver. Let $n(\Delta)$ be the number of vertices in $\mathcal{R}(\Delta)$ then $n(\Delta)$ is just the number of roots in the root system corresponding to Δ [4]. For example if Δ is of type $\mathbb{A}_5$ the root category $\mathcal{R}(\mathbb{A}_5)$ can be visualized as follows:

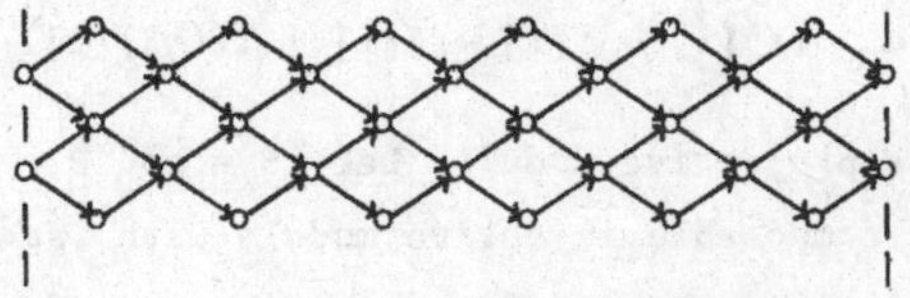

with identification along the dotted lines.

We recall the notion of a complete slice in $\mathcal{R}(\Delta)$.

First of all a set of vertices $X = \{x_1,\dots,x_r\}$ in $\mathbb{Z}\Delta$ is called a complete slice provided the following two conditions are satisfied:

(i) Given $x \in \mathbb{Z}\Delta$, then X contains precisely one point from the orbit $\{\tau^z x \mid z \in \mathbb{Z}\}$ of x.

(ii) If $y_o \to y_1 \to \dots \to y_s$ is a path in $\mathbb{Z}\Delta$, and y_o, y_s belong to X, then all y_i belong to X.

A set X of vertices of $R(\Delta)$ is called a complete slice provided $\pi^{-1}(x)$ restricted to a fundamental domain for the action of T^2 is a complete slice in $\mathbb{Z}\Delta$.

Next we want to transfer some concepts introduced in section 1 to $R(\Delta)$.

Let $x \in R(\Delta)$ and let $X^\cdot, Y^\cdot \in \pi^{-1}(x)$. Then clearly $\underline{\dim}\, X^\cdot = \underline{\dim}\, Y^\cdot$. Thus $\underline{\dim}\, x = \underline{\dim}\, X^\cdot$ does not depend on the choice of $X^\cdot$. It will be called the dimension vector of x.
Now let $x,y \in R(\Delta)$ and let $X^\cdot \in \pi^{-1}(x)$ and $Y^\cdot \in \pi^{-1}(y)$. Then we have defined a bilinear form on $\mathbb{Z}^n$, where n is the number of vertices of Δ, by $\langle X^\cdot, Y^\cdot\rangle = \sum_i (-1)^i \dim \mathrm{Hom}_{D^b(\mathrm{mod}\, k\vec{\Delta})}(X^\cdot, T^i Y^\cdot)$. This form is invariant under T^2.
Hence we may define $\langle x,y\rangle = \langle X^\cdot, Y^\cdot\rangle$ and obtain in this way a bilinear form on $R(\Delta)$. This bilinear form has the following interpretation. Let x be a vertex of $R(\Delta)$. We define a function $f_x : R(\Delta)_o \to \mathbb{Z}$ by $f_x(y) = \dim \mathrm{Hom}_{R(\Delta)}(x,y) - \dim \mathrm{Hom}_{R(\Delta)}(y, x)$. It is shown in [9] that $f_x(y) = \langle x,y\rangle$ and f_x is an additive function on $R(\Delta)$ (i. e. for all $y \in R(\Delta)_o$ we have $f_x(y) + f_x(\tau y) = \sum_{z \in y^-} f_x(z)$).

Let us add some additional notation. The symmetric bilinear form corresponding to $\langle -,-\rangle$ will be denoted by $(-,-)$. Finally we denote by $q_\Delta(x) = (x,x)$ the corresponding quadratic form. Observe that q_Δ is the usual Tits form for Δ [4].

Let $P_1,\dots,P_n$ be the indecomposable $k\vec{\Delta}$-modules considered as objects in $D^b(\mathrm{mod}\, k\vec{\Delta})$. Then $X = \{\pi P_1,\dots,\pi P_n\} = \{x_1,\dots,x_n\}$ is a complete slice in $R(\Delta)$. It will be called the standard complete slice. It is quite easy to see that for a vertex x we have $\underline{\dim}\, x = (f_{x_1}(x),\dots,f_{x_n}(x))$.

If we choose a different complete slice $Y = \{y_1,\dots,y_n\}$ we still may consider the vector $(f_{y_1}(y),\dots,f_{y_n}(x))$. It is well-known that different complete slices are obtained from each other by applying a sequence of reflections. These reflections may also be considered as linear transformations on $\mathbb{Z}^n$. Now let X, X' be two complete slices such that $X' = X^\sigma$ for some $\sigma = \sigma_1,\dots,\sigma_r$, with σ_i reflections. Denote again by σ the induced linear transformation on $\mathbb{Z}^n$. Then clearly $(f_{x_1}(x),\dots,f_{x_n}(x))\sigma = (f_{x_1'}(x),\dots,f_{x_n'}(x))$ for all vertices x. In particular we have the following lemma.

<u>Lemma</u>: Let $y_1,\ldots,y_s$ be vertices in $R(\Delta)$ and let X,X' be two complete slices in $R(\Delta)$. Then the vectors $(f_{x_1}(y_i),\ldots,f_{x_n}(y_i))$ for $1 \le i \le s$ are linearly independent if and only if the vectors $(f_{x_1'}(y_i),\ldots,f_{x_n'}(y_i))$ for $1 \le i \le s$ are linearly independent.

In the following we will always consider with a vertex x also its dimension vector $\underline{\dim}\, x$. Thus we may think of an underlying coordinate structure of x. So we may use the terminology that $x_1,\ldots,x_r$ are linearly independent, meaning that $\underline{\dim}\, x_1,\ldots,\underline{\dim}\, x_r$ are linearly independent.
Now we are ready to give the definition of a tilting set in $R(\Delta)$.

A subset $T = \{t_1,\ldots,t_n\}$ of vertices of $R(\Delta)$ is called a <u>tilting set</u> provided the following two conditions are satisfied:
(i) $\mathrm{Hom}_{R(\Delta)}(t_i,\tau t_j) = 0$ for $1 \le i,j \le n$
(ii) $t_1,\ldots,t_n$ form a $\mathbb{Z}$-basis.

Let $T = \{t_1,\ldots,t_n\}$ be a tilting set in $R(\Delta)$. Then we define the Cartan matrix $(C_T)_{ij} = (\dim \mathrm{Hom}_{R(\Delta)}(t_i,t_j))_{ij}$. Obviously $(C_T)_{ij} = (\langle t_i,t_j\rangle)_{ij}$ by the property (i) of a tilting set.

Following [7] we call a finite dimensional k-algebra A a <u>Δ-root algebra</u> or simply a root algebra if there exists a tilting set T in $R(\Delta)$ such that $A = (\mathrm{End}\, T)^{op}$.

Let us consider two examples of root algebras.

(a) $\Delta = \mathbb{A}_5$

We choose $T = \{t_1,\ldots,t_5\}$ as indicated.

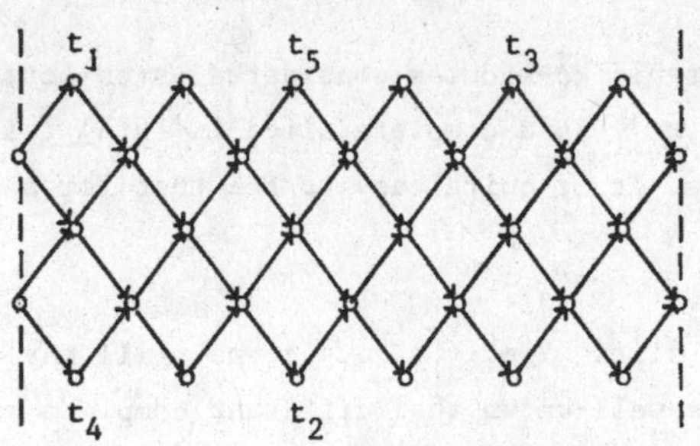

Then T is a tilting set in $R(\mathbb{A}_5)$ and $(\mathrm{End}\, T)^{op}$ is given by  .

(b) $\Delta = \mathbb{D}_4$

We choose $T = \{t_1,\ldots,t_4\}$ as indicated.

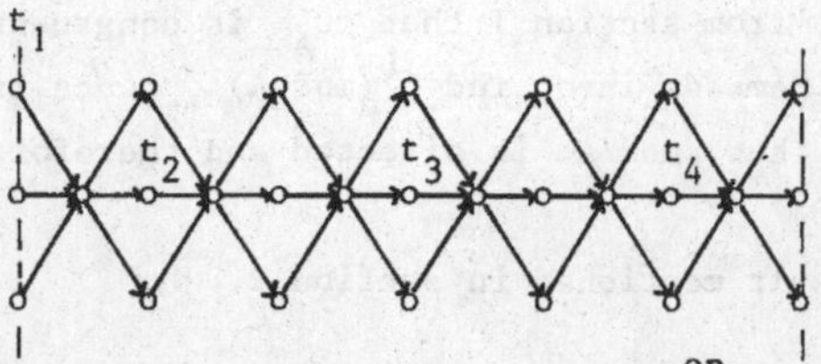

Then T is a tilting set in $R(\mathbb{D}_4)$ and $(\mathrm{End}\ T)^{op}$ is given by

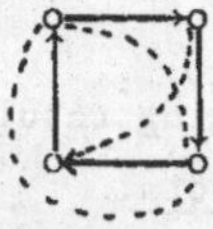

We call a tilting set T cycle-free, provided the ordinary quiver of $(\mathrm{End}\ T)^{op}$ does not contain an oriented cycle.

We quote from [7] the following result.

Theorem: Let T be a cycle-free tilting set in $R(\Delta)$. Then $A = (\mathrm{End}\ T)^{op}$ is an iterated tilted algebra.

Corollary: Let A be a Dynkin algebra. Then A is an iterated tilted algebra.

Proof. Let $P_1,\ldots,P_n$ be the indecomposable projective A-modules and let f be the congruence with $\langle x,y\rangle_A = \langle f(x),f(y)\rangle_\Delta$ for all $x,y \in \mathbb{Z}^n$. Let $t_i = f(\underline{\dim}\ P_i)$ for $1 \leq i \leq n$. Then it is quite easy to see [7] that $T = \{t_1,\ldots,t_n\}$ is a cycle-free tilting set. In view of the above the theorem it is cleary enough to show that $A \simeq (\mathrm{End}\ T)^{op}$. Since $B = (\mathrm{End}\ T)^{op}$ is an iterated tilted algebra it follows from section 1 that mod B is directed. In fact ind B embeds into $k(\mathbb{Z}\Delta)$. Moreover, we have $C_A = C_B$ by construction. Thus the assertion follows from the next lemma which is a special case of a result in [9].

Lemma: Let A,B be finite dimensional k-algebras with directed module category. If $C_A = C_B$ then $A \simeq B$.

3. Dynkin algebras

In this section we want to present the announced characterization of Dynkin algebras.

Theorem: Let A be a finite dimensional basic k-algebra and Δ a Dynkin diagram. Then the following are equivalent.

(i) $D^b(\mathrm{mod}\ A) \approx D^b(\mathrm{mod}\ k\vec{\Delta})$ as triangulated categories

(ii) A is a Dynkin algebra of type Δ

(iii) A is an iterated tilted algebra of type Δ

(iv) A is tilting equivalent to $k\vec{\Delta}$.

Proof. (i) $\Rightarrow$ (ii): It follows from section 1 that C_A is congruent to C_Δ. Moreover, we have that ind A embeds into ind $D^b(\text{mod } A)$, hence into ind $D^b(\text{mod } k\vec{\Delta})$. But ind $D^b(\text{mod } k\vec{\Delta}) \approx k(\mathbb{Z}\Delta)$. Thus mod A is directed and therefore A is a Dynkin algebra.

(ii) $\Rightarrow$ (iii): This is the result mentioned in section 2.

(iii) $\Rightarrow$ (iv): trivial.

(iv) $\Rightarrow$ (i): This is the invariance property with respect to tilting triples.

This finishes the proof of the theorem.

As an immediate application we obtain the following result.

Let $R_A = \{x \in \mathbb{Z}^n \mid q_A(x) = 1\}$ be the set of roots.

Corollary: Let A be a Dynkin algebra then dim induces a bijection from $\{\text{ind } D^b(\text{mod } A)\}/T^2$ to R_A.

Proof. This follows immediately from the results in section 1.

4. Examples

In this final section we want to give two examples of Dynkin algebras and compute the quiver of the derived category in these two situations.

(a) Let A be given by the following quiver with relations

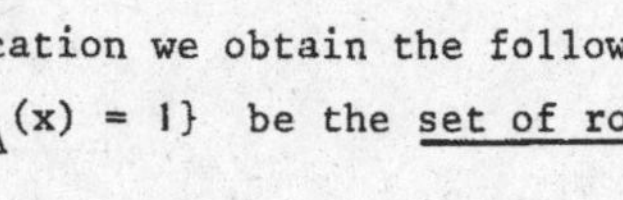

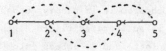

Then $\Gamma(A)$ has the following form

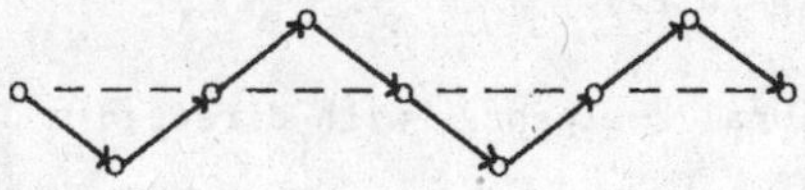

A is a Dynkin algebra of type $\mathbb{A}_5$.

Up to translation there are 15 isomorphism classes of indecomposable complexes in $D^b(\text{mod } A)$. In the following list the complexes $X^\cdot = (X^n, d^n)$ are normalized in such a way that $X^n = 0$ for $n > 0$ and $X^o \neq 0$.

$X_1^\cdot = \ldots 0 \to P(1) \to 0 \ldots$, $X_2^\cdot = \ldots 0 \to P(2) \to 0 \ldots$, $X_3^\cdot = \ldots 0 \to P(3) \to 0 \ldots$

$X_4^\cdot = \ldots 0 \to P(4) \to 0 \ldots$, $X_5^\cdot = \ldots 0 \to P(5) \to 0 \ldots$, $X_6^\cdot = \ldots 0 \to P(1) \to P(2) \to 0 \ldots$,

$X_7^\cdot = \ldots 0 \to P(2) \to P(3) \to 0 \ldots$, $X_8^\cdot = \ldots 0 \to P(3) \to P(4) \to 0 \ldots$,

$X_9^\cdot = \ldots 0 \to P(4) \to P(5) \to 0 \ldots$, $X_{10}^\cdot = \ldots 0 \to P(1) \to P(2) \to P(3) \to 0 \ldots$,

$X_{11}^\cdot = \ldots 0 \to P(2) \to P(3) \to P(4) \to 0 \ldots$, $X_{12}^\cdot = \ldots 0 \to P(3) \to P(4) \to P(5) \to 0 \ldots$,

$X^{\cdot}_{13} = \ldots\ 0 \to P(1) \to P(2) \to P(3) \to P(4) \to 0\ \ldots\ ,$

$X^{\cdot}_{14} = \ldots\ 0 \to P(2) \to P(3) \to P(4) \to P(5) \to 0\ \ldots\ ,$

$X^{\cdot}_{15} = \ldots\ 0 \to P(1) \to P(2) \to P(3) \to P(4) \to P(5) \to 0\ \ldots\ .$

The indecomposable A-modules correspond to $X^{\cdot}_1,\ldots,X^{\cdot}_6$, $X^{\cdot}_{10}$, $X^{\cdot}_{13}$ and $X^{\cdot}_{15}$.
The quiver of D^b(mod A) looks now as follows:

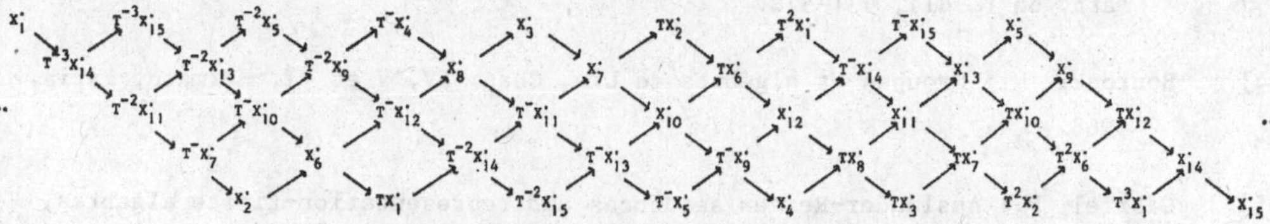

(b) Consider the algebra A given by the following quiver with relations

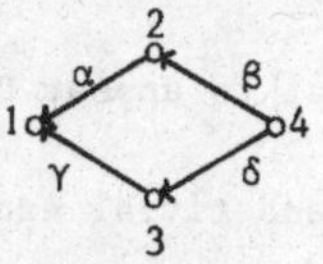

with $\alpha\beta = \gamma\delta$.

Then $\Gamma(A)$ has the following form

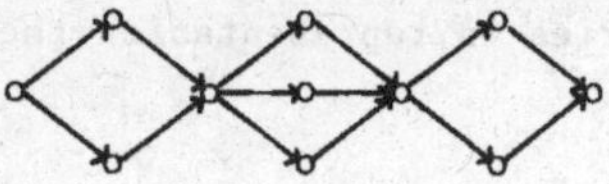

A is a Dynkin algebra of type $\mathbb{D}_4$.
Up to translation there are 12 isomorphism classes of indecomposable objects in D^b(mod A). In the following list these complexes are normalized as above.

$X^{\cdot}_1 = \ldots\ 0 \to P(1) \to 0\ \ldots\ ,\ X^{\cdot}_2 = \ldots\ 0 \to P(2) \to 0\ \ldots\ ,\ X^{\cdot}_3 = \ldots\ 0 \to P(3) \to 0\ \ldots\ ,$

$X^{\cdot}_4 = \ldots\ 0 \to P(4) \to 0\ \ldots\ ,\ X^{\cdot}_5 = \ldots\ 0 \to P(1) \to P(2) \to 0\ \ldots\ ,$

$X^{\cdot}_6 = \ldots\ 0 \to P(1) \to P(3) \to 0\ \ldots\ ,\ X^{\cdot}_7 = \ldots\ 0 \to P(2) \to P(4) \to 0\ \ldots\ ,$

$X^{\cdot}_8 = \ldots\ 0 \to P(3) \to P(4) \to 0\ \ldots\ ,\ X^{\cdot}_9 = \ldots\ 0 \to P(1) \to P(4) \to 0\ \ldots\ ,$

$X^{\cdot}_{10} = \ldots\ 0 \to P(2) \oplus P(3) \to P(4) \to 0\ \ldots\ ,\ X^{\cdot}_{11} = \ldots\ 0 \to P(1) \to P(2) \oplus P(3) \to 0\ \ldots\ ,$

$X^{\cdot}_{12} = \ldots\ 0 \to P(1) \to P(2) \oplus P(3) \to P(4) \to 0\ \ldots$

The indecomposable A-modules correspond to $X^{\cdot}_1,\ldots,X^{\cdot}_9$, $X^{\cdot}_{11}$ and $X^{\cdot}_{12}$.

The quiver of D^b(mod A) looks now as follows:

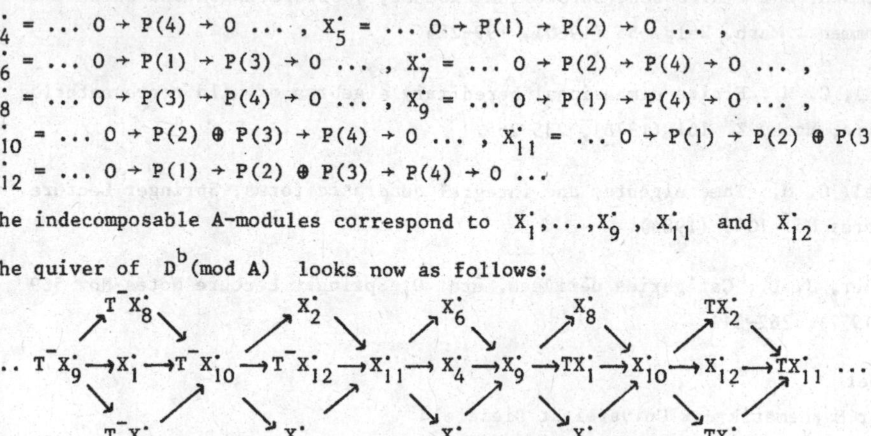

References

[1] Auslander, M.; Reiten, I.: Representation theory of Artin algebras III, VI, Comm. Algebra 3 (1975), 239-294, 5 (1977), 443-518.

[2] Bongartz, K.: Tilted algebras, Springer Lecture Notes No. 903 (1982), 26-38.

[3] Bongartz, K.; Gabriel, P.: Covering spaces in representation theory, Invent. Math. 65 (1981), 331-378.

[4] Bourbaki, N.: Groupes et algèbres de Lie, Chap. IV, V et VI, Hermann, Paris, 1968.

[5] Gabriel, P.: Auslander-Reiten sequences and representation-finite algebras, Springer Lecture Notes No. 831 (1980), 1-71.

[6] Gabriel, P.: The universal cover of a representation-finite algebra, Springer Lecture Notes, No. 903 (1982), 68-105.

[7] Happel, D.: Tilting sets on cylinders, to appear Proc. London Math. Soc.

[8] Happel, D.: Triangulated categories and tivial extension algebras, Proceedings ICRA IV, Ottawa 1984.

[9] Happel, D.: Triangulated categories in representation theory of finite dimensional algebras, to appear.

[10] Happel, D.; Ringel, C. M.: Tilted algebras, Trans. Amer. Math. Soc 274(2), (1982), 399-443.

[11] Happel, D.; Ringel, C. M.: The derived category of a tubular algebra, to appear.

[12] Riedtmann, Chr.: Algebren, Darstellungsköcher, Überlagerungen und zurück, Comment. Math. Helv. 55 (1980), 199-224.

[13] Ringel, C. M.: Finite dimensional hereditary algebras of wild representation type, Math. Z. 161 (1978), 235-255.

[14] Ringel. C. M.: Tame algebras and integral quadratic forms, Springer Lecture Notes No. 1099 (1984).

[15] Verdier, J. L.: Catégories dérivées, état 0, Springer Lecture Notes No. 569 (1977), 262-311.

Dieter Happel
Fakultät für Mathematik der Universität Bielefeld
D-4800 Bielefeld 1 / West Germany

DIMENSION FORMULAS RELATED TO A TAME QUIVER

Wim H. Hesselink

1. Introduction.

The purpose of this paper is to show how one can try to catch up with one of the main streams of mathematics, get lost in quiksand, and then discover unexpected facts.

So I start by reviewing some recent developments in pure mathematics, which induced me to construct two computer programs for the classification of the strictly upper triangular matrices. By comparison of the results, I found that matrices of order 7 with certain expected properties did not occur. In the effort to understand this fact I encountered the following result.

Theorem A. Let $m \leq 4$. Let $S_1, \ldots, S_m$ be finite dimensional subspaces of a linear space V. If x is an endomorphism of V, then

$$\sum_{i=2}^{m} \sum_{a=1}^{i-1} (-1)^{a+1} \dim(x^a S_{i-a} + S_i / S_i) \geq 0.$$

It was natural to ask whether this inequality might hold for an arbitrary number m, but for a long time I was unable to settle this question. Since the number of summands grows quadratically with m, I was glad to get a related result where the number of summands grow linearly.

Theorem C. Let S be a finite dimensional subspace of a linear space V. If x is an endomorphism of V with $x^4 S \subset S$, then

$$\dim(xS + S/S) - \dim(x^2 S + S/S) + \dim(x^3 S + S/S) \geq 0.$$

The chain of questions was broken by the emergence of a counter example. I found a configuration of five two-dimensional subspaces $S_1, \ldots, S_5$ in $\mathbb{R}^4$ such that

$$\sum_{i=2}^{5} \sum_{a=1}^{i-1} (-1)^{a+1} \dim(S_{i-a} + S_i / S_i) = -2.$$

It follows that the modification of theorem A is false for every number $m \geq 5$. The natural modifications of theorem C are also false. In a concluding remark I will discuss the relation with the tame quiver, mentioned in the title.

2. Some history.

In 1963, Konstant [7] showed that the set N of the nilpotent elements of a semisimple Lie algebra $\underline{g}$ is an irreducible normal variety and that the regular nilpotent elements form one conjugacy class which is dense in N. In 1966, Steinberg discussed in Moscow [13] various questions concerning N and the corresponding variety U of the unipotent elements of the adjoint group G. In 1968, Springer [11] proved that over a finite field the sets N and U have the same number of elements. To this end he introduced resolutions of N and U. If B is a Borel subgroup of G with Lie algebra $\underline{b}$ and nilradical $\underline{u}$, the resolution of N can be described as the projection morphism $p: Y \longrightarrow N$ where Y is the subvariety of $(G/B) \times N$ consisting of the pairs (gB,x) such that $x \in Ad(g)\underline{u}$. In 1970, Brieskorn [1] used Springer's resolution to show that N has rational surface singularities at the subregular points. So, the fibers $p^{-1}(x)$ of these points are Dynkin curves, cf. [8] and [14].

In the case $G = PGL(n)$, Spaltenstein [9] obtained in 1975 a parametrization of the irreducible components of all fibers $p^{-1}(x)$ by means of Young tableaux. In [3], I showed that the morphism $p: Y \longrightarrow N$ is a rational desingularization. Then Springer obtained the fundamental result [12], that the irreducible representations of the Weyl group W of G can be identified in the top cohomology of the fibers $p^{-1}(x)$. Note that the irreducible components of $p^{-1}(x)$ form a basis of this cohomology group. In the case of the classical groups, a parametrization of these components was obtained by Spaltenstein [10]. It also seemed interesting to know how these components are connected; that is to determine, say, the dimensions of the intersections of the irreducible components of $p^{-1}(x)$. In fact, the connectedness in codimension-one plays a role in the work of Kazhdan and Lusztig, cf. [6] 6.3. Since the fiber $p^{-1}(x)$ has a close resemblance to the intersection $\underline{u} \cap Ad(G)x$, it is a related question to find a finite stratification of $\underline{u}$ such that every stratum is irreducible, B-invariant, and contained in a single G-orbit.

If $G = PGL(n)$, then the nilradical $\underline{u}$ may be identified with the space of the strictly upper triangular matrices. In 1982, I proposed a finite classification for this space, cf. [4]. The classes are parametrized by certain combinatorial objects, called typrices. The classes are B-invariant, contained in a single G-orbit, but unfortunately, they need not be irreducible. In fact, arbitrarily wild varieties can occur. As another defect of the theory, let me point out that it is not clear whether the class corresponding to a given typrix is non-empty. I obtained a set of necessary conditions, but sufficiency was not claimed.

Around the same time, Bürgstein obtained the B-orbit classification for the strictly upper triangular matrices of order $n \leq 6$. If $n \leq 5$, all typrix classes turned out to be B-orbits. In the case $n = 6$, Bürgstein obtained 274 orbits and 1 one-parameter family of orbits, whereas my classification admitted 274 typrices. Working together in Groningen in the fall of 1983, Bürgstein and I formalized the orbit classification method and constructed a computer algorithm, cf. [2]. The program yields the B-orbits in $\underline{u}$ in the cases that G is simple of type $A_2,\dots,A_6,B_2,G_2,B_3,C_3$. It gives an almost complete classification in the cases A_7,B_4,C_4,D_4. The algorithm also works in the case of the B-action on the dual vector space $\underline{u}^*$.

Now that both classification methods were computerized, it was easy to compare the results. It turned out that all expected typrices of order 6 do occur (i.e. determine a non-empty class), and that only two members of the list of 1419 expected typrices of order 7 do not occur. The rest of the present paper contains an attempt to understand this non-occurence.

3. The non-occurring typrix

3.1. A typrix is a strictly upper triangular matrix of zeros and ones, of the same order as the matrices which it represents. Let C(A) denote the class of strictly upper triangular matrices corresponding to a given typrix A, cf. [4]. From now, let A be the typrix of diagram 1. It is one of the two expected but not occurring typrices of order 7. The other one is obtained from A by a reflection in the skew diagonal. So now the assertion is that the class C(A) is empty.

$$A = \begin{pmatrix} 0&0&1&0&1&0&0 \\ &0&0&0&0&1&0 \\ &&0&0&0&0&0 \\ &&&0&1&0&1 \\ &&&&0&0&0 \\ &&&&&0&1 \\ &&&&&&0 \end{pmatrix}$$

Diagram 1

In order to describe the set C(A) we need some notation. The dimension of a vector space W is denoted by $|W|$. If an endomorphism x induces an endomorphism of W, the induced rank is denoted by

$$|x : W| = |x(W)|.$$

Now, let V be the vector space K^7 with standard basis $e_1,\dots,e_7$. Let F_j be the subspace spanned by the vectors e_i with $i \le j$. So $F_0,\dots,F_7$ form the standard flag in V, and the space $\underline{u}$ of the strictly upper triangular matrices consists of the elements $x \in \mathrm{End}(V)$ with $x(F_j) \subset F_{j-1}$ for every index $j \ge 1$. It turns out that the typrix class C(A) is defined as the set of the elements $x \in \underline{u}$ satisfying

$$|x : F_2/0| = |x : F_4/F_1| = |x : F_6/F_4| = |x^3 : F_7/0| = 0,$$

$$|x : F_3/0| = |x : F_5/F_3| = |x : F_7/F_5| = |x : F_6/F_2| = 1,$$

$$|x^2 : F_5/0| = |x^2 : F_7/F_3| = 1, \text{ and } |x : F_6/F_1| = 2.$$

In other words, C(A) consists of the upper triangular matrices x of the following form satisfying the following conditions.

$$x = \begin{pmatrix} 0 & 0 & a & b & c & d & e \\ & 0 & 0 & 0 & f & g & h \\ & & 0 & 0 & i & j & k \\ & & & 0 & m & p & q \\ & & & & 0 & 0 & r \\ & & & & & 0 & s \\ & & & & & & 0 \end{pmatrix} \qquad \begin{array}{l} a \neq 0,\ m \neq 0,\ s \neq 0, \\ \begin{vmatrix} i & j \\ m & p \end{vmatrix} = 0 \neq \begin{vmatrix} f & g \\ m & p \end{vmatrix}, \\ ai + bm \neq 0 \neq mr + ps, \\ \text{and } x^3 = 0. \end{array}$$

So the entries 1 in the typrix A stand for inequalities, and some of the zeros stand for non-trivial equations. In particular, the top righthand entry $a_{17} = 0$ corresponds to the equation $x^3 = 0$. Now one should be able to verify and understand the outcome of the computer programs that the set C(A) is empty.

3.2. A direct matrix verification seems appropriate. So we calculate the matrix $y = x^3$. It turns out that all entries y_{ij} automatically vanish, except for the top righthand entry y_{17} which is equal to

$$y_{17} = a(ir + js) + b(mr + ps).$$

The condition $x^3 = 0$ is equivalent to $y_{17} = 0$. I shall show that this is incompatible with the other conditions. I only need the determinantal equality $ip = mj$ and the inequalities

$$ai + bm \neq 0 \neq mr + ps.$$

The product of these non-zero terms is non-zero. Applying the equality we get

$$0 \neq (ai + bm)(mr + ps) = my_{17}.$$

It follows that $y_{17} \neq 0$, and hence that $x^3 \neq 0$. So the conditions defining C(A) are incompatible. In other words, C(A) is empty.

3.3. This matrix calculation settles the fact, but it is not illuminating. Therefore after this calculation, I devised a more intrinsic proof.
We have the following data:

(1) $x^2F_7 \not\subset F_3$ and $F_2 \subset F_3$

(2) $x\, F_7 \subset F_6$

(3) $|x : F_6/F_2| \leq 1$

(4) $x^2F_5 \neq 0$ and $F_5 \subset F_6$

(5) $x F_2 = 0$

From these data we get

(6) $x^2F_7 \not\subset F_2$ (1)

(7) $x^2F_7 \subset x F_6$ (2)

(8) $x^2F_7 + F_2 = x F_6 + F_2$ (3,6,7)

(9) $x^2F_6 \neq 0$ (4)

(10) $x^3F_7 = x^2F_6 \neq 0$ (5,8,9)

Again the other data imply that $x^3 \neq 0$. This proves that the set C(A) is empty.

3.4. The reader may have the impression that this proof is not better than the first one. However, this proof shows that essentially only four subspaces play a role, namely

$$S_1 = F_7 , S_2 = F_6 , S_3 = F_2 , S_4 = F_o .$$

The crucial data are the inequalities

$$|x : S_1/S_2| \leq 0, \ |x : S_2/S_3| \leq 1, \ |x : S_3/S_4| \leq 0,$$

$$|x^2 : S_1/S_3| \geq 1, \ |x^2 : S_2/S_4| \geq 1,$$

and the conclusion is

$$|x^3 : S_1/S_4| \geq 1.$$

This pattern suggested the following general inequality:

$$|x^3 : S_1/S_4| - |x^2 : S_1/S_3| - |x^2 : S_2/S_4| + |x : S_1/S_2| + |x : S_2/S_3| + |x : S_3/S_4| \geq 0?$$

So now we leave our non-occurring typrix and the seven-dimensional space. We try to obtain understanding through generalisation.

4. Dimension inequalities.

4.1. The conjectural inequality of the end of the previous section is in fact a special case of our first main result.

Theorem A. Let $m \leq 4$. Let $S_1,\dots,S_m$ be finite dimensional subspaces of a linear space V. If $x \in \mathrm{End}(V)$, then

$$\sum_{i=2}^{m} \sum_{a=1}^{i-1} (-1)^{a+1} |x^a S_{i-a} + S_i / S_i| \geq 0.$$

Let A(m) denote the assertion in the case of m subspaces. So the theorem says that A(m) holds if $m \leq 4$. The assertions A(1) and A(2) are trivial. A(3) and A(4) can be proved directly. However, since I wanted to prove A(m) for all values of m, I tried inductive procedures in a more flexible setting. The powers of the endomorphism x are replaced by the compositions of homomorphisms in an infinite sequence of vector spaces. In fact, only finitely many spaces are non-zero in the applications I have in mind.

4.2. Consider a sequence of linear spaces V_i, $i \in \mathbb{Z}$, and homomorphisms $x_i : V_i \longrightarrow V_{i+1}$. When there is no danger of ambiguity, we write x instead of x_i. Similarly, the composite homomorphisms are denoted $x^a : V_{i-a} \longrightarrow V_i$. Let finite dimensional subspaces $S_i \subset V_i$ be given, of which only finitely many are non-zero.

Theorem B. Let $m \leq 4$. If $x^m S_{i-m} \subset S_m$ for all indices $i \in \mathbb{Z}$, then

$$\sum_{i \in \mathbb{Z}} \sum_{a=1}^{m-1} (-1)^{a+1} |x^a S_{i-a} + S_i / S_i| \geq 0.$$

Proof. If $m \leq 2$ the assertion is trivial. So we may assume that m is 3 or 4. Consider the partial sums

$$F_p = \sum_{i \leq p} \sum_{a=1}^{m-1} (-1)^{a+1} |x^a S_{i-a} + S_i / S_i|.$$

We claim that for all integers p we have

$$F_p \geq \Big| \sum_{i=0}^{m-1} x^i S_{p-i} / S_p + x^2 S_{p-2} \Big|.$$

This would prove the theorem. For, if p is sufficiently large, this formula reduces to the required inequality. On the other hand, if p is sufficiently small, then $S_i = 0$ for all indices $i \leq p$, so in that case the formula is trivial. Now the formula is proved by induction. So we may assume that

$$F_{p-1} \geq \Big| \sum_{i=0}^{m-1} x^i S_{p-1-i} / S_{p-1} + x^2 S_{p-3} \Big|.$$

If $A \subset B$ are subspaces of V_{p-1}, then

$$|B/A| \geq |xB + S_p/xA + S_p|.$$

Since $x^m S_{p-m} \subset S_p$, it follows that

$$F_{p-1} \geq |\sum_{i=0}^{m-1} x^i S_{p-i}/S_p + x S_{p-1} + x^3 S_{p-3}|.$$

So it remains to prove that

$$F_p - F_{p-1} \geq |S_p + xS_{p-1} + x^3 S_{p-3}/S_p| - |S_p + x^2 S_{p-2}/S_p|.$$

Since $m \in \{3,4\}$ and

$$F_p - F_{p-1} = \sum_{a=1}^{m-1} |x^a S_{i-a} + S_i/S_i|,$$

this is easy.If $m = 3$ one uses that $x^3 S_{p-3} \subset S_p$. If $m = 4$ one uses the general fact that

$$|A| + |B| \geq |A + B|.$$

4.3. Theorem A is an immediate consequence of theorem B. In fact, let the situation of theorem A be given. We form the infinite sequence of vectorspaces by putting $V_i = V$ if $1 \leq i \leq m$, and $V_i = 0$ otherwise. If $1 \leq i \leq m$, we consider the given subspace S_i as a subspace of V_i. Put $S_i = 0$ otherwise. If $1 \leq i \leq m$, then we use $x_i = x : V_i \longrightarrow V_{i+1}$. In all other cases we use the zero mapping. The condition $x^m S_{i-m} \subset S_m$ holds trivially. The conclusion of theorem B reduces to the conclusion of theorem A.

Slightly more general, if B(m) denotes the assertion of theorem B for a fixed value m, then B(m) implies A(m). In particular, a counterexample of A(m) induces a counterexample of B(m).

4.4. The distribution of the summands in theorem B is more homogeneous than in theorem A. This enables us to replace the double summation by a single summation.

Theorem C. Let S be a finite dimensional subspace of a linear space V. Let $m \leq 4$. If $x \in \mathrm{End}(V)$ satisfies $x^m S \subset S$, then

$$\sum_{a=1}^{m-1} (-1)^{a+1} |x^a S + S/S| \geq 0.$$

Let C(m) denote the assertion in the case of a fixed m. So the theorem says that C(m) holds if $m \leq 4$. Now we prove that the assertions B(m) and C(m) are equivalent.

First, assume that C(m) holds. Let $(V_i, x_i)_i$ be a sequence of spaces and homomorphisms. Let $V = \Sigma V_i$ denote the direct sum, and let $x \in \mathrm{End}(V)$ be the formal sum of the homomorphims $x_i : V_i \longrightarrow V_{i+1}$. Put $S = \Sigma S_i$. Then C(m) implies B(m).

Conversely, assume that B(m) holds. Let $S \subset V$ and $x \in \mathrm{End}(V)$ be given with $x^m S \subset S$. Consider a long sequence of n copies $V_1, \ldots, V_n$ of V connected by the homomorphisms $x : V_i \longrightarrow V_{i+1}$. Choose $S_i = S \subset V_i$. Add zero spaces V_i for all other indices i. Now the assertion B(m) implies that

$$B + (n - m + 1) \sum_{a=1}^{m-1} (-1)^{a+1} |x^a S + S/S| \geq 0,$$

where B is a constant independent of n. Letting n tend to infinity, we get C(m).

In particular this proves theorem C.

4.5. Finally, a counterexample was found after discussion with Van der Put. It is a counterexample to the assertion A(5), and the endomorphism x is the identity map. Let V be a four-dimensional space with basis vectors $e_1, \ldots, e_4$. Consider the two-dimensional subspaces $S_1, \ldots, S_5$ given by

$$S_1 = \langle e_1, e_2 \rangle, \; S_2 = \langle e_1, e_3 \rangle, \; S_3 = \langle e_1 + e_3, e_2 + e_4 \rangle,$$
$$S_4 = \langle e_2, e_4 \rangle, \; S_5 = \langle e_3, e_4 \rangle.$$

One easily verifies, that $|S_i + S_j/S_j| = 1$ if $j - i$ is odd, and that $|S_i + S_j/S_j| = 2$ if $i \neq j$ and $j - i$ is even. So in this case the alternating sum equals -2. This proves that assertion A(5) is false.

It is easy to see that counterexamples to A(5) induce counterexamples to A(m) for all numbers $m \geq 5$. So assertion A(m) is false for all numbers $m \geq 5$. By 4.3 and 4.4, it follows that B(m) and C(m) are also false for all numbers $m \geq 5$.

4.6. Remark. Theorem A is equivalent to a proposition concerning the representations of the tame quiver $\tilde{D}_5$, cf. [5]. This quiver consists of six nodes and five arrows as in the diagram. A representation consists of six finite dimensional vector spaces corresponding to the nodes, and five homomorphisms corresponding to the arrows. Let the rank of a homomorphism $f : V \longrightarrow W$ be denoted by $|f|$, let Kf denote the injection of the kernel of f into the domain V, let Cf denote the projection of the codomain W onto the cokernel W/f(V). The composition of homo-

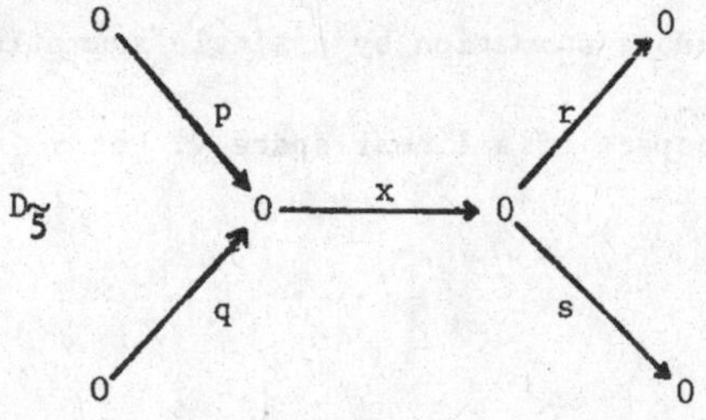

morphisms f and g is denoted by fg. Then theorem A is equivalent to the assertion that every representation of $\tilde{D}_5$ satisfies the inequality:

$$|(C_q)p| + |r(Ks)| + |r \times p| + |s \times q| - |r \times q| - |s \times p| \geq 0.$$

In fact, let $x' : V_o \longrightarrow V_1$ and $x : V_1 \longrightarrow V_2$ and $x'' : V_2 \longrightarrow V_3$ be homomophisms of vector spaces. Let subspaces $S_i \subset V_i$ be given, with factor spaces $W_i = V_i/S_i$. By consideration of the direct sum $V = \Sigma V_i$, one shows that theorem A is equivalent to

$$\sum_{i=1}^{3} \sum_{a=1}^{i-1} (-1)^{a+1} |S_{i-a} \longrightarrow W_i| \geq 0.$$

Now the proof of the equivalence with the quiver case is suggested by the names of the homomorphisms in the next diagram. In fact, in the quiver case one may assume that the homomorphism q is injective and that the homomorphism s is surjective.

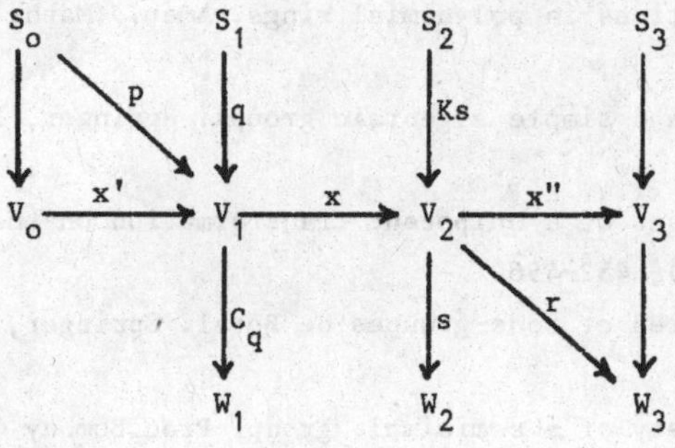

If one uses the known representation theory of $\tilde{D}_5$, cf. [5], then the inequality (and hence theorem A) becomes trivial. The theorems B and C however seem to be more difficult.

References.

1. E. Brieskorn: Singular elements of semisimple algebraic groups. Actes du Congrès International des Mathématiciens 1970, tome 2, p. 279-284.
2. H. Bürgstein, W.H. Hesselink: Algorithmic orbit classification for connected solvable groups. Preprint, Groningen, 1984.
3. W.H. Hesselink: Cohomology and the resolution of the nilpotent variety. Math. Ann. 223 (1976) 249-252.
4. W.H. Hesselink: A classification of the nilpotent triangular matrices. Compositio Math. (to appear).
5. V.G. Kac: Root systems, representations of quivers and invariant theory. Invariant theory, Proc. Montecatini 1982, p. 74-108. Springer (L.N.M. 996) 1983.
6. D.A. Kazhdan, G. Lusztig: Representations of Coxeter groups and Hecke algebras. Inventiones math. 53 (1979), 165-184.
7. B. Konstant: Lie group representations in polynomial rings. Amer.J.Math. 85 (1963) 327-404.
8. P. Slodowy: Simple singularities and simple algebraic groups. Springer, Berlin etc. 1980.
9. N. Spaltenstein: The fixed point set of a unipotent transformation on the flag manifold. Indag.Math. 38 (1976), 452-456.
10. N. Spaltenstein: Classes unipotentes et sous-groupes de Borel. Springer, Berlin etc. 1982.
11. T.A. Springer: The unipotent variety of a semisimple group. Proc.Bombay Colloq. Algebraic Geometry, 1968, p. 373-391.
12. T.A. Springer: Trigonometric sums, Green functions of finite groups and representations of Weyl groups. Inventiones math. 36 (1976), 173-207.
13. R. Steinberg: Classes of elements of semisimple algebraic groups. Proc. International Congress of Mathematics, Moscow 1966, p. 277-289.
14. R. Steinberg: Conjugacy classes in algebraic groups. Springer, Berlin etc. 1974.

Wim H. Hesselink
Mathematics Institute
Postbox 800
9700 AV GRONINGEN
The Netherlands

COHEN-MACAULAY MODULES OVER ISOLATED SINGULARITIES

Idun Reiten

Introduction

This note follows closely my lecture in Seminaire Malliavin in February 1985, and is based upon work with Maurice Auslander. Our aim is to explain how a technique first developed in the theory of finite dimensional algebras to show that there is only a finite number of indecomposable modules, can be extended to work also in the context of Cohen-Macaulay modules over commutative noetherian integrally closed Cohen-Macaulay local domains. This method is based upon the theory of almost split sequences, and we start out by giving the basic existence theorem for almost split sequences for finite dimensional algebras. We also explain the method in this context. It has been used extensively here, and has been further refined through the theory of coverings [6][7]. Also for lattices over classical orders we have existence of almost split sequences [1],[11], and a similar method was used for orders in [12]. We here explain how the method extends to Cohen-Macaulay modules over some commutative rings, and illustrate by sketching a proof of the fact that the 3-dimensional ring $R = \mathbb{C}[[X,Y,Z,U,V]]/(XZ-Y^2,XV-YU,YV-ZY)$ has only a finite number of indecomposable (maximal) Cohen-Macaulay modules, that is, R is of finite Cohen-Macaulay type. Here $\mathbb{C}$ denotes the complex numbers. In the last section we put the example into a more general context, by giving a survey of the main results known about when a commutative noetherian complete integrally closed Cohen-Macaulay local domain of dimension a least two is of finite Cohen-Macaulay type.

For background on almost split sequences for finite dimensional algebras we refer to [10].

1. The method using almost split sequences.

Let Λ be a finite dimensional algebra over a field k, and denote by mod Λ the category of finitely generated (left) Λ-modules. We recall the following [4].

<u>Definition</u> : An exact sequence $0 \to A \overset{f}{\to} B \overset{g}{\to} C \to 0$ in mod Λ is said to be almost split if (a) it is not split,

(b) A and C are indecomposable,

(c) given $h : X \to C$ with X indecomposable and h not an isomorphism, there is some $t : X \to B$ such that $gt = h$.

Denote by $D = \mathcal{H}om_k(\ ,k)$ the ordinary duality between mod Λ and mod Λ^{op}, and by Tr the transpose, that is, if $P_1 \to P_o \to C \to 0$ is a minimal projective presentation of C, then $\mathcal{H}om_\Lambda(P_o,\Lambda) \to \mathcal{H}om_\Lambda(P_1,\Lambda) \to TrC \to 0$ is exact. The basic theorem on almost split sequences is the following [3].

<u>Theorem 1</u> : Let Λ be a finite dimensional k-algebra. Given an indecomposable nonprojective C in mod Λ (or an indecomposable noninjective A), then there is an almost split sequence $0 \to A \to B \to C \to 0$ which is unique up to isomorphism. In this case we also have $A \simeq DTrC$.

We illustrate with the following example.

<u>Example 2</u> : Let Λ denote the k-algebra $\begin{pmatrix} k & o & o \\ o & k & o \\ k & k & k \end{pmatrix}$. Denote by P_i the indecomposable projective Λ-module given by the ith column of the matrix, and write $S_i = P_i/\underline{r}\,P_i$, where $\underline{r}$ denote the radical of Λ. Let I_i denote the injective envelope of S_i. Then it is well known that we have only the 6 indecomposable Λ-modules

$\mathcal{S} = \{P_1,P_2,P_3 = S_3,\ S_1 = I_1,\ S_2 = I_2,I_3\}$. It is then easy to see from the definition that we have the almost split sequences

$$0 \to P_1 \to I_3 \to S_2 \to 0$$
$$0 \to P_2 \to I_3 \to S_1 \to 0$$
$$0 \to S_3 \to P_1 \amalg P_2 \to I_3 \to 0$$

A method for proving that a finite dimensional k-algebra Λ is of finite representation type is based on the following [1].

<u>Theorem 3</u> : Let Λ be an indecomposable finite dimensional k-algebra, and assume that $\mathcal{S}$ is a finite nonempty set of indecomposable modules satisfying :

(a) If $0 \to A \to B \to C \to 0$ is almost split in mod Λ , and A or C is in $\mathcal{J}$, then each indecomposable summand of B is in $\mathcal{J}$.
(b) If P is an indecomposable projective in $\mathcal{J}$, then each indecomposable summand of $\underline{r}\,P$ is in $\mathcal{J}$.
(c) If I is an indecomposable injective in $\mathcal{J}$, then all indecomposable summands of $I/\mathrm{soc}\ I$ are in $\mathcal{J}$.

Then $\mathcal{J}$ consists of all indecomposable Λ-modules.

It is possible to apply this theorem to prove that a given algebra Λ has only a finite number of indecomposable modules, since there are criteria for deciding whether a given exact sequence is almost split without knowing all indecomposables. For example, we have the following [5].

<u>Lemma 4</u> : Let Λ be a finite dimensional k-algebra and C an indecomposable nonprojective Λ-module. If $0 \to \mathrm{DTr}C \to B \to C \to 0$ is a nonsplit exact sequence, and <u>End</u>(C), the endomorphism ring of C modulo the maps factoring through projectives is k, then this sequence is almost split.

We explain how we could use Theorem 3 to show that Λ in Example 2 has only a finite number of indecomposable modules. We start with the set $\mathcal{J}$, and form the three nonsplit exact sequences. Then we compute that $P_1 = \mathrm{DTr}\ S_2$, $P_2 = \mathrm{DTr}S_1$, $S_3 = \mathrm{DTr}I_3$, and $\underline{\mathrm{End}}(S_2) = k = \underline{\mathrm{End}}(S_1) = \underline{\mathrm{End}}(I_3)$. Hence it follows from Lemma 4 that the sequences are almost split, and Theorem 3 can be applied.

We shall not state the most general existence theorems for almost split sequences [1], but instead concentrate on the following case. Throughout the rest of the paper R will be a $\mathbb{C}$-algebra which is a commutative noetherian complete integrally closed Cohen-Macaulay local domain of dimension at least 2. We also assume $R/m \simeq \mathbb{C} \subset R$, where m denotes the maximal ideal of R. In this case $R \supset T$ where T is a regular local ring and R is a finitely generated T - module. Denote by $\mathcal{J}_R$ the category of finitely generated (maximal) Cohen-Macaulay R-modules, that is, the finitely generated R-modules which are free as T-modules. Denote by $w = \mathcal{H}om_T(R,T)$ the dualizing module. In this situation there is the following existence theorem [1][3].

<u>Theorem 5</u> : Let R be as above, and assume that T is an isolated singularity, that is R_p is regular local for each nonmaximal prime ideal p. If $C \neq R$ is indecomposable in $\mathcal{S}_R$(or $A \neq w$), then there is a unique almost split sequence $0 \to A \to B \to C \to 0$ in $\mathcal{S}_R$.

In this situation we have the following criterion for finite Cohen-Macaulay type, whose proof will appear elsewhere.

<u>Theorem 6</u> : Let R satisfy our standard assumptions, and assume R is an isolated singularity which is not Gorenstein. Assume that $\mathcal{S}$ is a finite nonempty set of indecomposable Cohen-Macaulay modules satisfying

(a) $R \in \mathcal{S}$

(b) If $0 \to A \to B \to C \to 0$ is almost split in $\mathcal{S}_R$ and A or C is in $\mathcal{S}$, then all indecomposable summands of $A \amalg B \amalg C$ are in $\mathcal{S}$.

Then $\mathcal{S}$ consists of all indecomposable Cohen-Macaulay modules.

2 - <u>Illustration on an example</u>.

Let now $R = \mathbb{C}[[X,Y,Z,U,V]]/(XZ-Y^2, XV-YU, YV-ZU)$, and we denote the image of X in R by x, etc. Then R is known to be a 3-dimensional ring satisfying our assumptions. The following ideals are known to be Cohen-Macaulay and this can also easily be checked directly : $R, A=(x,y)$, $A^2=(x^2,xy,y^2)$, $B=(x,y,u)$. Also in this case, when $0 \to E \to F \to G \to 0$ is almost split, there is a formula $E = \tau G$ giving the relationship between the end terms [1]. We do not explain the formula here, but mention that in our case we get $\tau A = D(\Omega^1 B)$, $\tau A^2 = B$, $\tau B = A^2$, $\tau(\Omega^1 B) = R$. Here $A = w$, D denotes the duality $\mathcal{H}om_R(\ ,w)$, and $\Omega^1 B$ denotes the first syzygy module for B.

We show that $\mathcal{S} = \{R, A, A^2, B, \Omega^1 B\}$ are all the indecomposable Cohen-Macaulay modules by using Theorem 6. We first show by direct construction that we have an exact sequence $0 \to R \to A \amalg A \amalg B \to \Omega^1 B \to 0$, which by dualizing gives an exact sequence $0 \to D(\Omega^1 B) \to R \amalg R \amalg A^2 \to A \to 0$, using that $D(B) = A^2$. Both are nonsplit exact sequences with "correct" end terms. We can show that $\underline{\mathrm{End}}(A) \simeq \mathbb{C}$, and use an analogue of Lemma 4 to conclude that the second sequence is almost split and duality to get that the first one also is almost split.

For the rest we use rank arguments, together with the important property of almost split sequences that if $F \neq w$, $G \neq R$, then F is a summand of the middle term with G on the right if and only if G is a summand of the middle term with F on the left [5]. If B is on the right, then $\tau B = A^2$ is on the left and applying this principle, R and A must be summands of the middle term. A rank argument then gives an almost split sequence $0 \to A^2 \to R \coprod A \to B \to 0$. Similarly, $\tau A^2 = B$ and $\Omega^1 B$ and $D(\Omega^1 B)$ must be in the middle when A^2 is on the right. Since the minimal number of generators for B is 3, $\Omega^1 B$ has rank 2. We then must have $\Omega^1 B \simeq D(\Omega^1 B)$, and an almost split sequence $0 \to B \to \Omega^1 B \to A^2 \to 0$. Applying Theorem 6 we see that $\mathcal{G}$ consists of all indecomposable Cohen-Macaulay modules.

A more detailed account of this proof will be given in a forthcoming publication with M. Auslander.

3. Known results about finite Cohen-Macaulay type.

In this section we give a survey on the main results known on when R has only a finite number of indecomposable Cohen-Macaulay modules, with R as before.

An interesting necessary condition is given by the following result of Auslander [3].

Theorem 7 : If $\mathcal{G}_R$ has almost split sequences, then R is an isolated singularity.

When $\dim R = 2$, then the Cohen-Macaulay modules coincide with the reflexive modules. In this case there is a complete description. A finite group $G \subset GL(n,\mathbb{C})$ induces a linear action on the power series ring $S = \mathbb{C}[[X_1,\ldots,X_n]]$, and we denote by $R = S^G$ the fixed ring under this action. Then we have the following.

Theorem 8 : If $\dim R = 2$, then R is of finite Cohen-Macaulay type if and only if $R \simeq \mathbb{C}[[X,Y]]^G$, where $G \subset GL(2,\mathbb{C})$ is a finite group.

That $R = \mathbb{C}[[X,Y]]^G$ is of finite type is due to Herzog [8], and the converse was proved by Artin-Verdier, Auslander [2], Esnault.

We point out that if $G \subset SL(2,\mathbb{C})$, then Auslander showed that the quiver built from almost split sequences for $R = \mathbb{C}[[X,Y]]^G$ and the resolution graph of the corresponding singularity determine each other, via the Mc Kay graph for $G \subset SL(2,\mathbb{C})$ [2].

If $G \subset SL(2,\mathbb{C})$, then $R = \mathbb{C}[[X,Y]]^G$ is known to be a hypersurface, of the form $R \simeq \mathbb{C}[[u,v,w]]/(f(u,v,w))$, where $f(u,v,w)$ is $u^{n+1} + v^2 + w^2$ or $u^{n-1} + uv^2 + w^2$ or $u^4 + v^3 + w^2$ or $u^3v + v^3 + w^2$ or $u^5 + v^3 + w^2$. Denote for $t \geq 0$ by f_t the polynomial in $t + 3$ variables $f_t(u,v,w,z_1,\ldots,z_t) = f(u,v,w) + z_1^2 + \ldots + z_t^2$. Then the simple hypersurface singularities over $\mathbb{C}$ in dimension at least two are the rings $\mathbb{C}[[u,v,w,z_1,\ldots,z_t]]/(f_t)$. When R satisfies our general assumptions, there is the following result on when a hypersurface is of finite Cohen-Macaulay type, showing interesting connections between the question of finite Cohen-Macaulay type and geometry.

<u>Theorem 9</u> : A hypersurface R is of finite Cohen-Macaulay type if and only if R is a simple hypersurface singularity.

That simple hypersurface singularities are of finite Cohen-Macaulay type was proved by Knörrer [9], and the converse was proved by Buchweitz, Greuel and Schreyer.

Moreover we have the following results whose proofs will appear in future publications with M. Auslander.

<u>Theorem 10</u> : A non regular ring $R = \mathbb{C}[[X_1,\ldots,X_n]]^G$, $n \geq 3$, is of finite Cohen-Macaulay type if and only if $R \simeq \mathbb{C}[[X_1,X_2,X_3]]^G$, with

$$G = < \begin{pmatrix} -1 & 0 & 0 \\ 0 & -1 & 0 \\ 0 & 0 & -1 \end{pmatrix} > \subset GL(3,\mathbb{C}).$$

Let $(n_1,\ldots,n_t)$ be a set of t positive integers. Consider the matrix

$$\begin{pmatrix} x_0^{(1)} & x_1^{(1)} & \ldots x_{n_1-1}^{(1)} \\ x_1^{(1)} & x_2^{(1)} & \ldots x_{n_1}^{(1)} \end{pmatrix} \ldots \begin{pmatrix} x_o^{(t)} & x_1^{(t)} & \ldots x_{n_t-1}^{(t)} \\ x_1^{(t)} & x_2^{(t)} & \ldots x_{n_t}^{(t)} \end{pmatrix}.$$

Let $S = \mathbb{C}[[x_o^{(1)},\ldots,x_{n_1}^{(1)},\ldots,x_o^{(t)},\ldots,x_{n_t}^{(t)}]]$, and denote by I the ideal generated by the 2×2 minors of this matrix. Then $R = S/I$ is by definition a scroll of type $(n_1,\ldots,n_t)$. R is known to satisfy our general assumption, and $\dim R = t+1$. When $t=1$, the scrolls are isomorphic to rings of the form $\mathbb{C}[[X,Y]]^G$, and are hence of finite Cohen-Macaulay type. For $t \geq 1$ we have the following result, and we are grateful to David Eisenbud and Finn Knudsen for many helpful conservations.

Theorem 11 : Let R be a scroll of type $(n_1,\ldots,n_t)$ with $t \geq 2$. Then T is of finite Cohen-Macaulay type if and only if R is of type (1,1) or (2,1).

We remark that a scroll of type (1,1) is just a simple hypersurface singularity. The scroll of type (2,1) is the example we computed in section 2. We point out that the other known R with dim $R \geq 3$ which is not an hypersurface and is of finite Cohen-Macaulay type, mentioned in Theorem 10, has 3 indecomposable Cohen-Macaulay modules. This can be proved for example by using the same method. It would be interesting to know if there are more than these two examples. Since Herzog has shown that if R is Gorenstein of finite Cohen-Macaulay type then R is a hypersurface [8], it is enough to consider rings which are not Gorenstein.

REFERENCES

[1] Auslander M. : Functors and morphisms determined by objects. Applications of morphisms determined by modules. Proc. Conf. Representation theory, Philadelphia 1976, 1-244 and 245-327, Marcel Dekker 1978.

[2] __________ : Rational singularities and almost split sequences. Trans. Amer. Math. Soc. (to appear).

[3] __________ : Isolated singularities and existence of almost split sequences. Proc. ICRA IV, Ottawa, Springer Lecture Notes (to appear).

[4] Auslander, M and Reiten, I : Representation theory of Artin algebras III : Almost split sequences, Comm. in Algebra , 3 (1975), 239-294.

[5] __________ : Representation theory of artin algebras IV : Comm. in Algebra 5 (1977), 443-518.

[6] Bongartz, K. and Gabriel, P : Covering spaces in representation theory, Invent. Math. 65 (1981/82), N°3, 331-378.

[7] Gabriel, P. : The universal cover of a representation-finite algebra, Proc. ICRA III (Puebla 1980), Springer Lecture Notes 903 (1981), 69-105.

[8] Herzog, J.: Ringe mit nur endlich vielen Isomorphieklassen von maximalen, unzerlegbaren Cohen-Macaulay-Modulen Math. Ann. 233 (1978), 21-34.

[9] Knörrer, H. : The Cohen-Macaulay modules over simple hypersurface singularities.

[10] Reiten, I : The use of almost split sequences in the representation theory of Artin algebras, Proc. ICRA III (Puebla 1980), Springer Lecture Notes 944 (1981), 29-104.

[11] Roggenkamp, K. and Schmidt, J. : Almost split sequences for integral group rings and orders, Comm. in Algebra 4 (1976), 893-917.

[12] Wiedemann, A : Orders with loops in their Auslander-Reiten graph, Comm. Algebra 9 (1981), N°6, 641-659.

SPIN-LIKE MODULES FOR CERTAIN INFINITE-DIMENSIONAL LIE ALGEBRAS

George B. Seligman*

Introduction

The ultimate aim of this paper is to develop an appropriate context for construction of integrable irreducible highest-weight modules for certain Kac-Moody algebras, especially those of types $B_\ell^{(1)}$, $C_\ell^{(1)}$, $D_\ell^{(1)}$. The constructions of the Kac-Moody modules will be given elsewhere, starting with the underlying spaces which are the modules M of this paper.

The paper may be read independently of the theory of Kac-Moody algebras, as concerning the Lie algebra $\mathcal{g}$ of transformations of a canonical infinite-dimensional space V, with a symplectic or symmetric bilinear form, generated by the maps

$$w \to (w,u)v \pm (w,v)u$$

where (w,u) is the form. By considering generalized even Clifford algebras of V, one is able to construct, and to deduce a number of properties of, a family of representations of $\mathcal{g}$ serving (with respect to a Borel subalgebra determined by a fixed basis) as an analogue of the finite-dimensional irreducible representations of the "$\mathcal{g}$" corresponding to finite-dimensional "V". In particular, the modules admit the action of a Casimir operator and, over $\mathbb{C}$, carry a "contravariant" positive-definite hermitian form.

1. Generalized Even Clifford Algebras.

We recall the general context in which we have considered analogues of the even Clifford algebra [4], [5], with some relaxation of constraints of finiteness of dimension. Namely, $\mathcal{D}$ is to be an involutorial division algebra of finite dimension over its center Z, a field of characteristic zero, and V a left $\mathcal{D}$-module with a nondegenerate form (u,v) which is hermitian $(\varepsilon = 1)$ or anti-hermitian $(\varepsilon = -1)$. This is to mean

$$(\alpha u,v) = \alpha(u,v);\ (v,u)^* = \varepsilon(u,v), \quad \text{where} \quad \alpha \in \mathcal{D};\ u,v \in V,$$

and $\alpha \to \alpha^*$ is the involution in $\mathcal{D}$. Let F be the field of elements of Z fixed by the involution.

---

*Research at Yale University and I.H.E.S., Bures-sur-Yvette, supported in part by National Science Foundation Grant No. MCS82-01333.

By setting $v \cdot \alpha = \alpha^* v$, V is made into a right $\mathcal{D}$-module V_r, so we may consider the F-vector space $V_r \otimes_{\mathcal{D}} V$; this is the quotient of $V \otimes_F V$ by the subspace generated by all

$$\alpha u \otimes v - u \otimes \alpha^* v .$$

The generalized even Clifford algebras are universal for maps $V_r \otimes_{\mathcal{D}} V$ into associative algebras with unit over F, the maps satisfying certain identities. Rather than working with $V_r \otimes_{\mathcal{D}} V$, we work with $V \otimes_F V$, letting the identity *i)* below carry the force of making our maps factor through $V \otimes_F V \to V_r \otimes_{\mathcal{D}} V$.

Specifically, let k be a non-negative integer, and let (A_k, φ) be universal for associative F-algebras A with unit and F-linear maps $\varphi' : V \otimes_F V \to A$, satisfying the following identities:

i) $\varphi'(\alpha u \otimes v) = \varphi'(u \otimes \alpha^* v)$;

ii) $\varphi'(u \otimes v) + \varepsilon\varphi'(v \otimes u) = \frac{k}{2}\,(t((v,u) + (v,u)^*))1$;

iii) $[\varphi'(u \otimes v), \varphi'(w \otimes x)] = \varphi'(uS_{w,x} \otimes v) + \varphi'(u \otimes vS_{w,x})$;

iv) $\varphi'_{k+1} = 0$, where the F-multilinear map $\varphi'_t : V^{2t} \to A$ is defined inductively as follows:

$$\varphi'_1(u_1; v_1) = \varphi'(u_1 \otimes v_1);$$

$$\varphi'_{t+1}(u_1,\ldots,u_{t+1}; v_1,\ldots,v_{t+1}) =$$

$$\sum_{i=1}^{t+1} \varphi'(u_i \otimes v_1)\varphi'_t(u_1,\ldots,\hat{u}_i,\ldots,u_{t+1}; v_2,\ldots,v_{t+1})$$

$$+ \sum_{i<j} \varphi'_t(u_1,\ldots,[u_i,u_j,v_1],\ldots,\hat{u}_j,\ldots,u_t; v_2,\ldots,v_{t+1}).$$

Here the notations are as follows: For $\alpha \in \mathcal{D}$, $t(\alpha) \in Z$ is the unique element of Z such that $\alpha = t(\alpha)1 + \alpha_0, \alpha_0 \in [DD]$;

$$uS_{w,x} = (u,w)x - \varepsilon(u,x)w;$$

$$[u,v,w] = uS_{v,w} - \varepsilon(v,w)u.$$

Then $\varphi(V \otimes V)$ generates A_k, and the argument for finiteness of dimension in [4] applies here to show that A_k is <u>locally</u> finite. The following identities are proved by induction on t; for $t = 1$, they are respectively *iii)* and *i)* above.

(I) For each t, $[\varphi_t(u_1,\ldots,u_t;v_1,\ldots,v_t),\varphi(u \otimes v)]$

$$= \sum_{i=1}^{t} \varphi_t(u_1,\ldots,u_i S_{u,v},\ldots,u_t;v_1,\ldots,v_t)$$

$$+ \sum_{i=1}^{t} \varphi_t(u_1,\ldots,u_t;v_1,\ldots,v_i S_{u,v},\ldots,v_t).$$

(II) For each $\lambda \in \mathcal{D}$ and each t,

$$\sum_{i=1}^{t} \varphi_t(u_1,\ldots,\lambda u_i,\ldots,u_t;v_1,\ldots,v_t)$$

$$= \sum_{i=1}^{t} \varphi_t(u_1,\ldots,u_t;v_1,\ldots,\lambda^* v_i,\ldots,v_t).$$

Now (I) and (II) may be invoked to prove the following:

Proposition 1. There is an involutorial F-antiautomorphism of A_k mapping $\varphi(u \otimes v)$ to $\varepsilon\varphi(v \otimes u)$.

The idea of the proof is as follows: Let A_k^{op} be the same vector space as A_k, with the same unit and reversed multiplication, so that the (set-theoretic) identity map is an anti-isomorphism between A_k and A_k^{op}. Then it is easy to verify that $\psi: V \otimes_F V \to A_k^{op}$ defined by $\psi(u \otimes v) = \varepsilon\varphi(u \otimes v)$ satisfies *i*) – *iii*). For *iv*), one shows inductively on t that

(III) $$\psi_t(u_1,\ldots,u_t;v_1,\ldots,v_t) = \varepsilon^t \varphi_t(v_t,\ldots,v_1;u_t,\ldots,u_1).$$

Once (III) is verified, then $\psi_{k+1} = 0$, and we have a unique homomorphism $\eta: A_k \to A_k^{op}$ of F-algebras making commutative the diagram

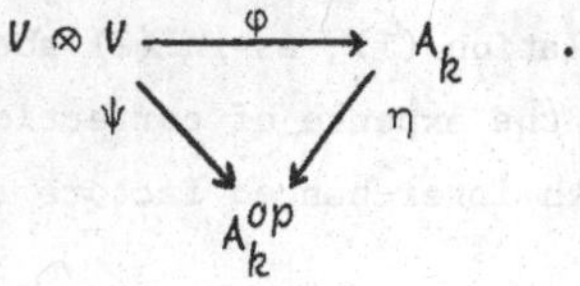

Working set-theoretically we see that once that $\eta^2(\varphi(u \otimes v)) = \varphi(u \otimes v)$ for all u,v, and that $\eta^2: A_k \to A_k$ is a homomorphism. Because $\varphi(V \otimes V)$ generates A_k, it follows that η, viewed as map $A_k \to A_k$, is an involution.

The relation (III) requires a rather careful induction. For $t = 2$, we have

(1) $$\psi_2(u_1,u_2;v_1,v_2) = \varphi(v_2 \otimes u_2)\varphi(v_1 \otimes u_1) + \varphi(v_2 \otimes u_1)\varphi(v_1 \otimes u_2)$$

$$+ \varepsilon\varphi(v_2 \otimes [u_1,u_2,v_1]),$$

the multiplications on the right being those of A_k. Now use *iii*) to write the second term on the right as

$$(2)\quad \varphi(v_1 \otimes u_2)\varphi(v_2 \otimes u_1) - \varphi(v_1 S_{v_2,u_1} \otimes u_2) - \varphi(v_1 \otimes u_2 S_{v_2,u_1}) .$$

Now expand the last term of (1) and the last two of (2), using $(u,v) = \varepsilon(v,u)^*$ and *ii*), to show that these terms combine to form simply $\varphi([v_2,v_1,u_2] \otimes u_1)$. Substituting in (1) yields (III) for $t = 2$.

When $t > 2$, we proceed by induction. The definition, induction and expansion give [assuming (III) for lower values]:

$$(3)\quad \psi_t(u_1,\ldots,u_t;v_1,\ldots,v_t) = \varepsilon^t\varphi_{t-1}(v_t,\ldots,v_2;u_{t-1},\ldots,u_1)\varphi(v_1 \otimes u_t)$$
$$+ \varepsilon^t \sum_{i=1}^{t-1}\sum_{j=2}^{t} \varphi(v_j \otimes u_t)\varphi_{t-2}(v_t,\ldots\hat{v}_j,\ldots,v_2;u_{t-1},\ldots,\hat{u}_i,\ldots,u_1)\varphi(v_1 \otimes u_i)$$
$$+ \varepsilon^{t-1}\sum_{i=1}^{t-1}\sum_{2\le j<\ell} \varphi_{t-2}(v_t,\ldots,[v_\ell,v_j,u_t],\ldots,\hat{v}_j,\ldots,v_2;u_{t-1},\ldots$$
$$,\hat{u}_i,\ldots,u_1)\varphi(v_1 \otimes u_i)$$
$$+ \varepsilon^{t-1}\sum_{j=2}^{t}\sum_{i<m<t} \varphi(v_j \otimes u_t)\varphi_{t-2}(v_t,\ldots,\hat{v}_j,\ldots,u_2;u_{t-1},\ldots,\hat{u}_m,\ldots$$
$$,[u_i,u_m,v_1],\ldots,u_1)$$
$$+ \varepsilon^{t-1}\sum_{2\le j<\ell}\sum_{i<m<t} \varphi_{t-2}(v_t,\ldots,[v_\ell,v_j,u_t],\ldots,\hat{v}_j,\ldots,v_2;u_{t-1},\ldots,\hat{u}_m,$$
$$\ldots,[u_i,u_m,v_1],\ldots,u_1)$$
$$+ \varepsilon^{t-1}\sum_{i<t}\varphi_{t-1}(v_t,\ldots,v_2;u_{t-1},\ldots,[u_i,u_t,v_1],\ldots,u_1).$$

Now we use the commutation-relation (I), as in (2) above, to interchange the factors in the first term of (3), at the expense of correction-terms as in the right-hand side of (I). The term with interchanged factors combines with the second and fourth terms of (3) to give

$$(4)\quad \sum_{j=1}^{t} \varepsilon^t\varphi(v_j \otimes u_t)\varphi_{t-1}(v_t,\ldots,\hat{v}_j,\ldots,v_2;u_{t-1},\ldots,u_1).$$

The third and fifth terms of (3) combine as above to

(5) $$\varepsilon \sum_{2\le j<\ell} \psi_{t-1}(u_1,\ldots,u_{t-1};v_2,\ldots,\hat{v}_j,\ldots,[v_\ell,v_j,u_t],\ldots,v_t)$$
$$= \varepsilon^t \sum_{2\le j<\ell} \varphi_{t-1}(v_t,\ldots,[v_\ell,v_j,u_t],\ldots,\hat{v}_j,\ldots,v_2;u_{t-1},\ldots,u_1).$$

The sixth term of (3) and the terms of the transformed first term involving members $\ldots u_i S_{v_1,u_t}\ldots$ combine to $-\varepsilon^t \sum_{i=1}^{t-1} \varphi_{t-1}(v_t,\ldots,v_2;u_{t-1},\ldots,(u_t,v_1)u_i,\ldots,u_1)$, which by (II) is equal to

$$-\varepsilon^{t-1} \sum_{j=2}^{t} \varphi_{t-1}(v_t,\ldots,(v_1,u_t)v_j,\ldots,v_2;u_{t-1},\ldots,u_1),$$

and this combines with the remaining terms of the transformed first term of (3) to give

(6) $$\varepsilon^t \sum_{j=2}^{t} \varphi_{t-1}(v_t,\ldots,[v_j,v_1,u_t],\ldots,v_2;u_{t-1},\ldots,u_1).$$

Thus (3) = (4) + (5) + (6), and the sum of these last three is, by definition of φ_t,

$$\varepsilon^t \varphi_t(v_t,\ldots,v_1;u_t,\ldots,u_1).$$

This completes the proof by induction.

2. Canonical Bases, Local Finiteness and Irreducible Modules.

Hereafter the division algebra $\mathcal{D}$ is assumed to be F. Thus "$\varepsilon = 1$" means the form is symmetric (and bilinear), while for $\varepsilon = -1$ it is skew-symmetric. We further assume that V has (finite or) countable dimension and that the form is split. For $\varepsilon = 1$, this is taken to mean that either V has a basis

$$\ldots,e_{-\ell},\ldots,e_{-1},e_1,e_2,\ldots,e_\ell,\ldots \qquad (\text{"even" case})$$

or a basis

$$\ldots,e_{-\ell},\ldots,e_{-1},e_0,e_1,\ldots,e_\ell,\ldots \qquad (\text{"odd" case})$$

with $(e_i,e_j) = \delta_{i,-j}$ for all i,j. For $\varepsilon = -1$, there is to be a basis as in the even case, with $(e_i,e_j) = (\operatorname{sgn} j)\delta_{i,-j}$.

Let V_∞ denote a fixed split infinite-dimensional space satisfying the above conditions, and let $n_1 < n_2 < n_3 < \ldots$ be an infinite strictly increasing sequence of positive integers. (In the even case, with $\varepsilon = 1$, we assume $n_1 \ge 3$ in order to avoid degeneracies. Let V_r be the subspace of V_∞ with basis those e_j with $|j| \le n_r$ for $r = 1,2,\ldots$.

For fixed k, and for $r < s$, the injection $V_r \to V_s$ yields a composite map $V_r \otimes V_r \to V_s \otimes V_s \xrightarrow{\varphi} A_k(V_s)$, where $\varphi = \varphi^{(s)}$ is the canonical map of our construction. It follows that there is an algebra-homomorphism $A_k(V_r) \to A_k(V_s)$ mapping $\varphi^{(r)}(u \otimes v)$ to $\varphi^{(s)}(u \otimes v)$ for $u,v \in V_r$. If $\varepsilon = 1$, let k be arbitrary; if $\varepsilon = -1$, let k be even. (If k is odd and $\varepsilon = -1$, each $A_k(V_r) = \{0\}$, and it follows that $A_k(V_\infty) = \{0\}$ - see [4]. [5].) Then if $g(V_r)$ is the Lie algebra of F-endomorphisms of V_r skew with respect to the form, we identify $g(V_r)$ with the subalgebra of $g(V_s)$, for $r < s$, stabilizing V_r and annihilating those e_j with $n_r < |j| \le n_s$. That is, $g(V_r)$ is the subspace (necessarily a subalgebra) of $g(V_s)$ generated by the $S_{u,v}(u,v \in V_r)$. We know from [4] that $A_k(V_r)$ is a semi-simple associative algebra, containing and generated by $g(V_r)$, a Lie subalgebra via a mapping sending $S_{u,v}(u,v \in V_r)$ to $\varphi(u \otimes v) - \frac{k}{2}(v,u)1$. We have also described all the irreducible $A_k(V_r)$-modules as irreducible $g(V_r)$-modules. In terms of the injection $g(V_r) \to g(V_s)$, which makes the diagram

$$\begin{array}{ccc} g(V_r) & \hookrightarrow & g(V_s) \\ \downarrow & & \downarrow \\ A_k(V_r) & \hookrightarrow & A_k(V_s) \end{array}$$

commutative, it is an easy matter to see that, as one runs through the irreducible $A_k(V_s)$-modules, regarded as $g(V_s)$-modules and then as $g(V_r)$-modules by restriction, all the irreducible $A_k(V_r)$-modules occur as irreducible $g(V_r)$-modules in such restrictions. By the semisimplicity of $A_k(V_r)$, it follows that the map $A_k(V_r) \to A_k(V_s)$ is <u>injective</u>. Thus $A_k(V_\infty)$ may be regarded as the (increasing) union of the $A_k(V_r)$, $1 < r < \infty$.

For $i \ge 1$, let $T_i = \varepsilon S_{e_{-i},e_i} \in g(V_r)$ for every r with $i \le n_r$;

Let $X_{\alpha_{i+1}} = S_{e_{-i},e_{i+1}}$, $X_{-\alpha_{i+1}} = S_{e_i,e_{-(i+1)}}$, in $g(V_r)$ whenever $i+1 \le n_r$. If $\varepsilon = -1$, let $X_{\alpha_1} = S_{e_1,e_1}$, $X_{-\alpha_1} = S_{e_{-1},e_{-1}}$; if $\varepsilon = 1$, in the even case, let $X_{\alpha_1} = S_{e_1,e_2}$, $X_{-\alpha_1} = S_{e_{-1},e_{-2}}$, and in the odd case $X_{\alpha_1} = S_{e_0,e_1}$, $X_{-\alpha_1} = S_{e_{-1},e_0}$ - these are in <u>every</u> $g(V_r)$. Then the $T_i \in g(V_r)$ span a splitting Cartan subalgebra of $g(V_r)$, call it H_r. The $X_{\alpha_j} \in g(V_r)$ are root-vectors for a set of simple roots $\alpha_1,\ldots,\alpha_{n_r}$ relative to H_r, with the $X_{-\alpha_j}$ belonging to the negatives of these roots. The corresponding "coroots" $H_1,\ldots,H_{n_r}$ are given by $H_j = T_j - T_{j-1}$

if $j > 1$, with $H_1 = T_1$ if $\varepsilon = -1$; $H_1 = T_1 + T_2$, if $\varepsilon = 1$, in the even case; $H_1 = 2T_1$ in the odd case with $\varepsilon = 1$. The algebra $g(V_r)$ is simple, with Dynkin diagram as follows:

C_{n_r} :

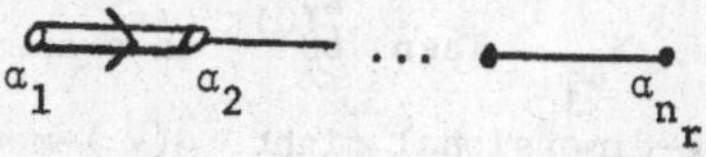

($\varepsilon = -1$)

D_{n_r} :

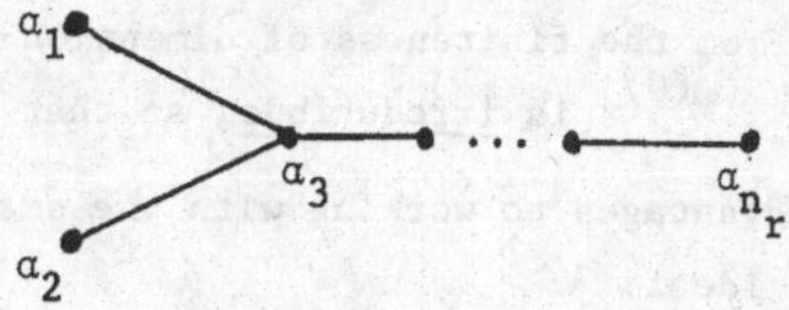

($\varepsilon = 1$, even case)

B_{n_r} :

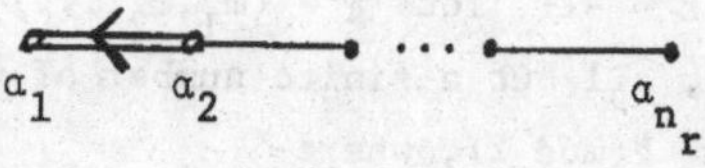

($\varepsilon = 1$, odd case)

Let $\lambda_1, \lambda_2, \ldots, \lambda_{n_r}$ be the corresponding fundamental weights, i.e., a basis for the dual space of H_r dual to $H_1, \ldots, H_{n_r}$. The highest weights of the irreducible $g(V_r)$-modules arising as $A_k(V_r)$-modules are of the form $\lambda = \sum_{i=1}^{n_r} m_i \lambda_i$, m_i non-negaitve integers satisfying $s(\lambda) \leq k$, $s(\lambda) \equiv k \pmod 2$, where:

$$s(\lambda) = 2\Sigma m_i \qquad \text{if} \quad \varepsilon = -1;$$

$$s(\lambda) = m_1 + 2 \sum_{i>1} m_i \qquad \text{if} \quad \varepsilon = 1, \text{ odd case};$$

$$s(\lambda) = m_1 + m_2 + 2 \sum_{i>2} m_i \qquad \text{if} \quad \varepsilon = 1, \text{ even case, [5].}$$

Fix one such λ, satisfying these conditions. The irreducible <u>right</u> $g(V_r)$-module with this as highest weight may be given a presentation as follows: There is a generator v_λ satisfying $v_\lambda H_i = m_i v_\lambda$ for all i, $v_\lambda X_{\alpha_i} = 0$ for all i, $v_\lambda X_{-\alpha_i}^{m_i+1} = 0$ for all i. Identifying the listed elements of $g(V_r)$ with their images in $A_k(V_r)$, we see that the annihilator in $A_k(V_r)$ is a right ideal containing all $H_i - m_i 1$, all X_{α_i} and all $(X_{-\alpha_i})^{m_i+1}$. Because v_λ generates an irreducible right $A_k(V_r)$-module, its annihilator is a <u>maximal</u> right ideal in $A_k(V_r)$.

Conversely, we define b_λ to be the right ideal in $A_k(V_r)$ generated by the $H_i - m_i 1$, the X_{α_i} and the $X_{-\alpha_i}^{m_i+1}$, and let M_λ be the quotient-module $A_k(V_r)/b_\lambda$, necessarily different from zero by the last paragraph. The right $g(V_r)$-module M_λ is thus generated by $\bar{1} (\neq 0)$, the image of $1 \in A_k(V_r)$, and $\bar{1}$

is annihilated by the elements above. Thus, by [1]. Ex. 15, p. 232, M_λ is the irreducible $\mathfrak{g}(V_r)$-module of highest weight λ, so is $A_k(V_r)$-irreducible, and thus $\mathfrak{b}_\lambda$ is a <u>maximal right ideal</u> in $A_k(V_r)$.

Actually this assertion can be sharpened. Let $\mathfrak{b}_\lambda^{(0)}$ be the right ideal in $A_k(V_r)$ generated by the $H_i - m_i 1$ and the X_{α_j}. Then $\mathfrak{b}_\lambda^{(0)}$ is proper by the above, and $M_\lambda^{(0)} = A_k(V_r)/\mathfrak{b}_\lambda^{(0)}$ is a finite-dimensional right $\mathfrak{g}(V_r)$-module generated by $\bar{1}$, a highest weight-vector. From the finiteness of dimension and the representations of $\mathfrak{sl}(2)$ it follows that $M_\lambda^{(0)}$ is <u>irreducible</u>, so that $\mathfrak{b}_\lambda^{(0)} = \mathfrak{b}_\lambda$. There will be some technical advantages to working with the smaller generating set above for our maximal right ideal.

Still with k fixed (and even if $\varepsilon = -1$), let $\Lambda = (m_1, m_2, \ldots)$ be an infinite sequence of non-negative integers, all but a finite number of which are zero, and which satisfy $s(\Lambda) \le k$, $s(\Lambda) \equiv k \pmod 2$, where

$$s(\Lambda) = \begin{cases} 2(m_1 + m_2 + \ldots), & \text{if } \varepsilon = -1; \\ m_1 + m_2 + 2(m_3 + m_4 + \ldots), & \text{if } \varepsilon = 1, \text{ even case;} \\ m_1 + 2(m_2 + m_3 + \ldots), & \text{if } \varepsilon = 1, \text{ odd case.} \end{cases}$$

With H_i and X_{α_i} as above, now for <u>all</u> positive integral values of i, let $\mathfrak{b}_\Lambda$ be the right ideal in $A_k = A_k(V_\infty)$ generated by all $H_i - m_i 1$ and all X_{α_i}. Then $\mathfrak{b}_\Lambda$ <u>is a (proper) maximal right ideal in</u> A_k.

This assertion is an easy consequence of the facts in the finite-dimensional case. To see that the right ideal is proper, note that otherwise we should have a relation $1 = \sum_{j=1}^{s} b_j c_j$ in A_k, where the b_j are among our generators, and $c_j \in A_k$. But then all members of the relation are in $A_k(V_r)$ for sufficiently large r, where the right ideal generated by the b_j is proper. To see that $\mathfrak{b}$ is maximal, let $y \notin \mathfrak{b}_\Lambda$, so that $y \in A_k(V_r)$ for some r, but $y \notin \mathfrak{b}_\lambda^{(0)}$, in the notations above. Then $\mathfrak{b}_\lambda^{(0)} + yA_k(V_r) = A_k(V_r)$, so that $1 \in \mathfrak{b}_\Lambda + yA_k$. The result is

<u>Proposition 2: Let V_∞ be a split F-space as at the beginning of this section. Fix a non-negative integer k, k even if $\varepsilon = -1$. Then for each infinite sequence $\Lambda = \{m_i\}$ as above with $s(\Lambda) \le k$, $s(\Lambda) \equiv k \pmod 2$, the right quotient module of $A_k = A_k(V_\infty)$ by the right ideal generated by all $H_i - m_i 1$ and all X_{α_i}, for $i = 1,2,\ldots$ is an irreducible module M_Λ. The generator $\bar{1}$ of M_Λ, the image of $1 \in A_k$, is annihilated by $X_{-\alpha_i}^{m_i+1}$ for each i, but by no lower power of $X_{-\alpha_i}$.</u>

(The last assertion, about "no lower power", is immediate from the theory of finite-dimensional modules for $\mathfrak{sl}(2)$.)

The statement of Proposition 2 is completely analogous to our assertions about $A_k(V_r)$ and $b_\Lambda^{(0)}$, where the Lie algebra $g(V_r)$ was involved. In the case of Proposition 2, the role of $g(V_r)$ is played by the Lie algebras of skew endomorphisms T of V_∞ such that for some $n = n(T)$, $e_j T = 0$ for all $|j| > n$. We denote this Lie algebra by $g^{(0)}(V_\infty)$, and note that it is simply the set of linear combinations of all $S_{u,v}$ $(u,v \in V_\infty)$. In the next section, we develop more analogies with the finite-dimensional case.

3. Elementary Properties.

We fix k, Λ and M_Λ as in the last section, and we denote by v_Λ the canonical image in M_Λ of $1 \in A_k = A_k(V_\infty)$. We note the following facts:

a) v_Λ is annihilated by all $\varphi(e_i \otimes e_j)$ for $i+j > 0$.

b) v_Λ is an eigenvector for each $\varphi(e_j \otimes e_{-j})$, $j \geq 0$, the corresponding eigenvalue being

$$\frac{k}{2} - b_1 - b_2 - \ldots - b_j, \quad \text{if } \varepsilon = -1;$$

$$\left.\begin{cases} \frac{k}{2} - \frac{1}{2}b_1 - \frac{1}{2}b_2 - b_3 - b_4 - \ldots - b_j, \ j > 1 \\ \frac{k}{2} - \frac{1}{2}b_1 + \frac{1}{2}b_2, \ j = 1 \end{cases}\right\} \varepsilon = 1, \text{ even case;}$$

$$\frac{k}{2} - \frac{1}{2}b_1 - b_2 - \ldots - b_j, \quad \text{if } \varepsilon = 1, \text{ odd case.}$$

c) If v' is any weight vector in M_Λ (i.e., $v'H_i = \mu(H_i)v'$ for all H_i, where $\mu(H_i) \in F$) and if $v'\varphi(e_i \otimes e_j) = 0$ for all $i+j > 0$, then the subspace M'_Λ of M_Λ spanned by all

(7) $$v'X_{-\alpha_{i_1}} \ldots X_{-\alpha_{i_t}} = v''$$

for all finite sequences $(i_1, \ldots, i_t)$ of positive integers is an A_k-submodule, so is either M_Λ or $\{0\}$. Thus, with $v' = v_\Lambda$, M_Λ is spanned by weight vectors.

d) Each element v''((7) above) is a common eigenvector for all $\varphi(e_j \otimes e_{-j})$, $j \geq 0$, or is zero. Each (nonzero) weight vector v is associated with a sequence $c_0, c_1, \ldots$ of rational numbers such that $v\varphi(e_j \otimes e_{-j}) = c_j v$, and the values of c_j stabilize for j sufficiently large, at the same value for every weight vector. We call the sequence $c_0, c_1, c_2, \ldots$ the <u>weight</u> of v.

e) In sequences $\{c_j\}$ as above, we may introduce a linear ordering: $\{c_j\} > \{d_j\}$ if the largest j for which $c_j \neq d_j$ has $c_j < d_j$. Then if v'' of (7) is a weight vector and if $i > 0$, $v''X_{-\alpha_i}$ is either 0 or a weight vector, whose weight is strictly less than that of v''.

f) Those M_Λ for distinct Λ satisfying the conditions above are nonisomorphic A_k-modules, and the intersection, for all such Λ, of the ideals $\mathrm{Ann}_{A_k}(M_\Lambda)$ in A_k is $\{0\}$. Thus A_k is a semi-primitive associative F-algebra.

g) For each weight, the totality of weight-vectors of this weight is a finite-dimensional subspace of M_Λ; this dimension is 1 if the weight is that of v_Λ.

A few remarks to substantiate these assertions:

a) In fact, all such $\varphi(e_i \otimes e_j)$ are in the Lie subalgebra of A_k generated by the X_{α_i}: If $i,j > 0$, with $j-i > 0$, then

$$[\ldots[[\varphi(e_{-i} \otimes e_{i+1}), (e_{-(i+1)} \otimes e_{i+2})]\varphi(e_{-(i+2)} \otimes e_{(i+3)}]\ldots\varphi(e_{-(j-1)} \otimes e_j)]$$

$$= \pm\,\varphi(e_{-i} \otimes e_j),$$

and this is $\varphi(e_j \otimes e_{-i})$, except for sign. If $0 \leq i \leq j$, and $\varepsilon = -1$,

$$[[\ldots[\varphi(e_1 \otimes e_1), \varphi(e_{-1} \otimes e_2)]\ldots]\varphi(e_{-(j-1)} \otimes e_j)] = \pm\, 2\varphi(e_1 \otimes e_j),$$

and then

$$[[\ldots[\varphi(e_1 \otimes e_j), \varphi(e_{-1} \otimes e_2]\ldots]\varphi(e_{-(i-1)} \otimes e_i)] = \pm\,\varphi(e_i \otimes e_j).$$

Similar calculations apply in the other cases,

In b), we have $\varphi(e_j \otimes e_{-j}) = k - \varepsilon\varphi(e_{-j} \otimes e_j)$, while $T_j = \varepsilon S_{e_{-j},e_j} = \varepsilon\varphi(e_{-j} \otimes e_j) - \frac{k}{2}$, in our identification of $g^{(0)}$ with a Lie subalgebra of A_k. Thus

$$\varphi(e_j \otimes e_{-j}) = \frac{k}{2} - T_j$$

for each j,

$$H_j = T_j - T_{j-1} = \varphi(e_{j-1} \otimes e_{-(j-1)}) - \varphi(e_j \otimes e_{-j}),$$

and $\varphi(e_{j-1} \otimes e_{-(j-1)}) - \varphi(e_j \otimes e_{-j})$ multiples v_Λ by b_j. The fact that $H_1 = T_1$, $T_1 + T_2$, or $2T_1$ in our respective cases then enables us to complete the calculation. [$\varphi(e_0 \otimes e_0) = \frac{k}{2}1$, when e_0 is present.]

In c), we use the fact from a) that the subspace $g^{(0)}$ is generated as Lie algebra by the $X_{\pm\alpha_i}$ and the $\varphi(e_j \otimes e_{-j})$. Then the fact that each $[X_{-\alpha_i}, S_{e_j,e_{-j}}]$ is a scalar multiple of $X_{-\alpha_i}$ shows that M'_Λ is stable under the action of all $S_{e_j,e_{-j}}$ and all $X_{-\alpha_i}$. We further have $[X_{-\alpha_i}, X_{\alpha_j}] = 0$ unless $i = j$, in which case the bracket is a multiple of H_i. It follows that M'_Λ is stable under the action of $g^{(0)}$, hence is an A_k-submodule. When $v' = v_\Lambda$, we know $M'_\Lambda = M_\Lambda$, and the displayed linear generators v'' are weight-vectors, by the calculation indicated in this argument.

In d), the first assertion follows as in b). When our weight vector is v_Λ, we see that the assertion about the sequence holds, the sequence $c_0, c_1, \ldots$ being $\frac{k}{2}, \frac{k}{2} - \frac{1}{2}b_1, \frac{k}{2} - \frac{1}{2}b_1 - b_2, \ldots$ in the third case, and stabilizing at $\frac{k}{2} - \frac{1}{2}b_1 - (b_2 + b_3 + \ldots) \geq 0$ as soon as $\{b_j\}$ stablizes at 0. The other cases are similar. Then $[X_{-\alpha_{i+1}}, \varphi(e_j \otimes e_{-j})]$ is typically an element

$$[\varphi(e_{-(i+1)} \otimes e_i), \varphi(e_j \otimes e_{-j})] = (-\delta_{ij} + \delta_{i+1,j})\varphi(e_{-(i+1)} \otimes e_i),$$

so that operating with $X_{-\alpha_{i+1}}$ on a weight vector of weight $\{c_\nu\}$ produces 0 or a weight vector of weight $(c_0, \ldots, c_i - 1, c_{i+1} + 1, c_{i+2}, \ldots)$. For $X_{-\alpha_1}$ in place of $X_{-\alpha_{i+1}}$, the corresponding shift is to $(c_0, c_1 + 2, c_2, \ldots)$ if $\varepsilon = -1$; to $(c_0, c_1 + 1, c_2 + 1, c_3, \ldots)$ if $\varepsilon = 1$, even case; to $(c_0, c_1 + 1, c_2, c_3, \ldots)$ if $\varepsilon = 1$, odd case. [We may drop c_0 or set $c_0 = \frac{k}{2}$ in all cases; it is immaterial.]

This assertion d) now follows from c) and induction on t. [The common stable value for all c_i is always the same, and the same as given above for v_Λ.]

e) now follows from the calculations of d).

To see f), note that an isomorphism of M_Λ with $M_{\Lambda'}$ would map v_Λ to a weight vector annihilated by all X_{α_i}. If this weight vector is not a multiple of v'_Λ it would generate a submodule as in c) of $M_{\Lambda'}$, all of whose weights are less than its weight, which in turn is strictly less than that of $v_{\Lambda'}$. Thus Λ' could not be a weight of this submodule, and we have a contradiction. Thus the isomorphism maps v_Λ to a multiple of $v_{\Lambda'}$, and hence $\Lambda = \Lambda'$ by the action of the H_i.

The second assertion follows because any non-zero $a \in A_k$ is in some $A_k(V_n)$, a finite-dimensional semisimple algebra, so fails to annihilate some $A_k(V_n)/b_\lambda^{(0)}$. By taking Λ as a trivial extension of λ, we see that the $g(V_n)$-

module generated by a highest weight vector for A_k/b_Λ is isomorphic to $A_k(V_n)/b_\lambda^{(0)}$, so is not annihilated by a.

To see g), let $v' = v_\Lambda$ in (7). If $t \neq 0$, the weight of v'' is strictly less than that of v_Λ, so the second assertion holds. Now suppose $\{c_\nu\}$ is a given weight of M_Λ; we claim that $\{c_\nu\}$ determines the sequence $i_1,\ldots,i_t$ of (7), up to order; thus t is determined and the corresponding weight-space has dimension at most $t!$. All this requires is the linear independence of the α_i as functions on the set $\{\varphi(e_j \otimes e_{-j})\}$, and this last is easily verified.

4. Casimir Operators.

For each algebra $g(V_r)$, the trace form $\mathrm{Tr}(XY)$ is nondegenerate and symmetric on $g(V_r)$, and there is a positive rational constant b (depending on r, as well as on ε and on evenness-oddness if $\varepsilon = 1$) such that $B(X,Y) = b\mathrm{Tr}(XY)$ is the Killing form of $g(V_r)$ (whenever $\dim V_r \geq 5$, so that $g(V_r)$ is simple.) Recall that the Casimir element of a semisimple Lie algebra g is obtained by taking an arbitrary basis $\{X_i\}$ for g, $\{Y_i\}$ the corresponding dual basis with respect to the Killing form, and forming the element $C = \sum_i X_iY_i$ in the universal enveloping algebra. The action of C is a g-endomorphism of each g-module. We compute C in our finite sections.

$\varepsilon = -1$: Let $e_{-\ell},\ldots,e_{-1},e_1,\ldots,e_\ell$ be a basis for a finite-dimensional symplectic V as above. Then the S_{e_i,e_j} for $i \leq j$ form a basis for $g = g(V)$, with dual basis relative to $\mathrm{Tr}(XY)$ as follows:

$$S_{e_{-j},e_j} = S_{e_j,e_{-j}} \leftrightarrow \tfrac{1}{2}S_{e_j,e_{-j}}, \quad 1 \leq j \leq \ell;$$

$$S_{e_j,e_j} \leftrightarrow -\tfrac{1}{4}S_{e_{-j},e_{-j}}, \quad 1 \leq |j| \leq \ell;$$

$$S_{e_i,e_j} \leftrightarrow -\tfrac{1}{2}S_{e_{-j},e_{-i}}, \quad 0 < i < j \leq \ell;$$

$$S_{e_{-i},e_j} \leftrightarrow \tfrac{1}{2}S_{e_{-j},e_i}, \quad 0 < i < j \leq \ell.$$

Thus the Casimir element is

$$(8) \qquad C = b^{-1}\{\tfrac{1}{2}\sum_{j=1}^{\ell} S^2_{e_j,e_{-j}} - \tfrac{1}{4}\sum_{j=1}^{\ell}(S_{e_j,e_j}S_{e_{-j},e_{-j}} + S_{e_{-j},e_{-j}}S_{e_j,e_j})$$
$$- \tfrac{1}{2}\sum_{0<i<j}(S_{e_i,e_j}S_{e_{-j},e_{-i}} + S_{e_{-j},e_{-i}}S_{e_i,e_j}) + \tfrac{1}{2}\sum_{0<i<j}(S_{e_{-i},e_j}S_{e_{-j},e_i}$$
$$+ S_{e_{-j},e_i}S_{e_{-i},e_j})\}.$$

For fixed k, the image in $A_k(V)$ of bC is thus $\overline{bC}$ =

$$(9)\qquad \frac{1}{2}\sum_{j=1}^{\ell}(\varphi(e_j\otimes e_{-j})-\frac{k}{2}1)^2-\frac{1}{2}\sum_{j=1}^{\ell}\varphi(e_j\otimes e_j)\varphi(e_{-j}\otimes e_{-j})$$

$$-\sum_{j=1}^{\ell}(\varphi(e_j\otimes e_{-j})-\frac{k}{2}1)-\sum_{0<i<j}\varphi(e_i\otimes e_j)\varphi(e_{-j}\otimes e_{-i})$$

$$-\frac{1}{2}\sum_{0<i<j}(\varphi(e_i\otimes e_{-i})+\varphi(e_j\otimes e_{-j})-k1)+\sum_{0<i<j}\varphi(e_{-i}\otimes e_j)\varphi(e_{-j}\otimes e_i)$$

$$+\frac{1}{2}\sum_{0<i<j}(\varphi(e_i\otimes e_{-i})-\varphi(e_j\otimes e_{-j})),$$

making use of $S_{u,v}\to\varphi(u\otimes v)-\frac{k}{2}(v,u)1$ and the identities *ii*), *iii*).

A little simplification leads to the following form:

$$(10)\qquad \overline{bC}=\frac{1}{2}\sum_{j=1}^{\ell}\varphi(e_j\otimes e_{-j})^2-\frac{1}{2}\sum_{j=1}^{\ell}\varphi(e_j\otimes e_{-j})$$

$$-\sum_{j=1}^{\ell}j\varphi(e_j\otimes e_{-j})-\frac{1}{2}\sum_{j=1}^{\ell}\varphi(e_j\otimes e_j)\varphi(e_{-j}\otimes e_{-j})$$

$$-\sum_{0<i<j}\varphi(e_i\otimes e_j)\varphi(e_{-j}\otimes e_{-i})+\sum_{0<i<j}\varphi(e_{-i}\otimes e_j)\varphi(e_{-j}\otimes e_i)$$

$$+\frac{k^2\ell}{8}1+\frac{k}{2}1+\frac{k}{2}\binom{\ell}{2}1$$

or $\overline{bC}=C'+(\frac{k^2\ell}{8}+\frac{k}{2}\binom{\ell+1}{2})1$, where C' is the sum of the non-scalar terms in our expression (10) for $\overline{bC}$. We shall refer to C' as <u>the</u> $A_k(V)$-<u>renormalized</u> <u>Casimir operator</u>.

When $\varepsilon = 1$ and we are in the even case, we have

$$(11)\qquad C'=\frac{1}{2}\sum_{j=1}^{\ell}\varphi(e_j\otimes e_{-j})^2-\frac{k}{2}\sum_{j=1}^{\ell}\varphi(e_j\otimes e_{-j})-\sum_{j=1}^{\ell}(j-1)\varphi(e_j\otimes e_{-j})$$

$$+\sum_{0<i<j}\varphi(e_i\otimes e_j)\varphi(e_{-j}\otimes e_{-i})+\sum_{0<i<j}\varphi(e_{-i}\otimes e_j)\varphi(e_{-j}\otimes e_i),$$

$$\overline{bC}=C'+\frac{k^2\ell}{8}1+\frac{k}{2}\binom{\ell}{2}1.$$

When $\varepsilon = 1$ and the case is odd,

(12) $$C' = \frac{1}{2}\sum_{j=1}^{\ell}\varphi(e_j \otimes e_{-j})^2 - \frac{k+1}{2}\sum_{j=1}^{\ell}(e_j \otimes e_{-j}) - \sum_{j=1}^{\ell}(j-1)\varphi(e_j \otimes e_{-j})$$

$$+ \sum_{0\le i\le j}\varphi(e_i \otimes e_j)\varphi(e_{-j} \otimes e_{-i}) + \sum_{0<i<j}\varphi(e_{-i} \otimes e_j)\varphi(e_{-j} \otimes e_i),$$

$$\overline{bC} = C' + \frac{k^2\ell}{8}1 + \frac{k\ell^2}{4}1.$$

If M is an irreducible right $A_k(V)$-module, whose highest weight as $g(V)$-module is λ, and if $v_\lambda\varphi(e_j \otimes e_{-j}) = c_j v_\lambda$ for each j, where v_λ is a highest-weight vector, we have, for ρ as described below,

$$v_\lambda\overline{bC} = b(\lambda + 2\rho|\lambda)v_\lambda,$$

where $(\mu|\nu)$ denotes a positive-definite rational-valued symmetric bilinear form on the rational span of the roots relative to our fixed split maximal torus, induced by the restriction of the Killing form to this torus. (See [3], p. 247, or [2], pp. 51 ff.) Furthermore, if μ is any weight of M, $\mu \neq \lambda$, we have

$$(\lambda + 2\rho|\lambda) > (\mu + 2\rho|\mu).$$

(See [3], p. 246, or [2], Ch. 1, 10), or [1], p. 126).

Now let $C^{(0)}$ be the sum of the first three members (those involving terms with factors $\varphi(e_j \otimes e_{-j})$) in our respective expressions of C' in (10), (11), (12). The remaining terms annihilate v_λ, so that

$$b(\lambda + 2\rho|\lambda)v_\lambda = v_\lambda\overline{bC} = v_\lambda C^{(0)} + \frac{k^2\ell}{8}v_\lambda + \begin{cases} \frac{k}{2}\binom{\ell+1}{2}v_\lambda & (10) \\ \frac{k}{2}\binom{\ell}{2}v_\lambda & (11) \\ \frac{k}{2}\frac{\ell^2}{2}v_\lambda & (12), \end{cases}$$

according to the separate cases, or

(14) $$b(\lambda + 2\rho|\lambda) = \frac{1}{2}\sum_{j=1}^{\ell}c_j^2 - \frac{k}{2}\sum_{j=1}^{\ell}c_j - \sum_{j=1}^{\ell}jc_j + \frac{k^2\ell}{8} + \frac{k}{2}\binom{\ell+1}{2};$$

$$b(\lambda + 2\rho|\lambda) = \frac{1}{2}\sum_{j=1}^{\ell}c_j^2 - \frac{k}{2}\sum_{j=1}^{\ell}c_j - \sum_{j=1}^{\ell}(j-1)c_j + \frac{k^2\ell}{8} + \frac{k}{2}\binom{\ell}{2};$$

$$b(\lambda + 2\rho|\lambda) = \frac{1}{2}\sum_{j=1}^{\ell}c_j^2 - \frac{k+1}{2}\sum_{j=1}^{\ell}c_j - \sum_{j=1}^{\ell}(j-1)c_j + \frac{k^2\ell}{8} + \frac{k\ell^2}{4},$$

according to the cases as previously ordered.

The inner product $(\mu|\nu)$ may be described as follows: The $S_{e_j,e_{-j}}$, $1 \le j \le \ell$, and the $\frac{1}{2b}S_{e_j,e_{-j}}$ are dual bases for our splitting maximal torus. Then $(\mu|\nu) = \sum_{j=1}^{\ell} \mu(S_{e_j,e_{-j}})\nu(\frac{1}{2b}S_{e_j,e_{-j}}) = \frac{1}{2b}\sum_{j=1}^{\ell} \mu(S_{e_j,e_{-j}})\nu(S_{e_j,e_{-j}})$.

The element ρ is characterized by having the value 1 at each of our fundamental set $H_1,\ldots,H_\ell$ of "coroots". In the three cases, these are, respectively:

(17) $$-S_{e_1,e_{-1}}, S_{e_1,e_{-1}} - S_{e_2,e_{-2}},\ldots,S_{e_{\ell-1},e_{-(\ell-1)}} - S_{e_\ell,e_{-\ell}};$$

(18) $$-S_{e_1,e_{-1}} - S_{e_2,e_{-2}}, S_{e_1,e_{-1}} - S_{e_2,e_{-2}},\ldots,S_{e_{\ell-1},e_{-(\ell-1)}} - S_{e_\ell,e_{-\ell}};$$

(19) $$-2S_{e_1,e_{-1}}, S_{e_1,e_{-1}} - S_{e_2,e_{-2}},\ldots,S_{e_{\ell-1},e_{-(\ell-1)}} - S_{e_\ell,e_{-\ell}}.$$

The values of ρ are thus given by, respectively,

(20) $$\rho(S_{e_1,e_{-1}}) = -1, \rho(S_{e_2,e_{-2}}) = -2,\ldots,\rho(S_{e_\ell,e_{-\ell}}) = -\ell;$$

(21) $$\rho(S_{e_1,e_{-1}}) = 0, \rho(S_{e_2,e_{-2}}) = -1, \rho(S_{e_3,e_{-3}}) = -2,\ldots,\rho(S_{e_\ell,e_{-\ell}}) = -(\ell-1);$$

(22) $$\rho(S_{e_1,e_{-1}}) = -\frac{1}{2}, \rho(S_{e_2,e_{-2}}) = -\frac{3}{2},\ldots,\rho(S_{e_\ell,e_{-\ell}}) = -\frac{1}{2} - (\ell-1).$$

Thus for μ as above, we have, respectively,

(23) $$(\mu|2\rho) = -\frac{1}{b}\sum_{j=1}^{\ell} j\mu(S_{e_j,e_{-j}});$$

(24) $$(\mu|2\rho) = -\frac{1}{b}\sum_{j=1}^{\ell} (j-1)\mu(S_{e_j,e_{-j}});$$

(25) $$(\mu|2\rho) = -\frac{1}{2b}\sum_{j=1}^{\ell} \mu(S_{e_j,e_{-j}}) - \frac{1}{b}\sum_{j=1}^{\ell} (j-1)\,(S_{e_j,e_{-j}}).$$

Now suppose μ is any weight $\neq \lambda$ of our irreducible representation of highest weight λ. Let v_μ be a non-zero weight vector of weight μ, so that $v_\mu\varphi(e_j \otimes e_{-j}) = d_j v_\mu$ for each j. (Thus the sequence $(d_0,d_1,\ldots,d_\ell,0,\ldots)$ is the weight of v_μ in our previous sense.) From the above, we compute

$(\mu + 2\rho|\mu)$ in terms of the d_i: In all cases,

$$b(\mu|\mu) = \frac{1}{2} \sum_{j=1}^{\ell} \mu(S_{e_j,e_{-j}})^2 = \frac{1}{2} \sum_{j=1}^{\ell} (d_j - \frac{k}{2})^2$$

$$= \frac{1}{2} \sum_{j=1}^{\ell} d_j^2 - \frac{k}{2} \sum_{j=1}^{\ell} d_j + \frac{k^2\ell}{8}.$$

When $\varepsilon = -1$, $b(2\rho|\mu) = -\sum_{j=1}^{\ell} jd_j + \frac{k}{2}\binom{k+1}{2}$; when $\varepsilon = 1$ and "e_0" is absent,

$$b(2\rho|\mu) = -\sum_{j=1}^{\ell} (j-1)d_j + \frac{k}{2}\binom{\ell}{2};$$

when $\varepsilon = 1$ and e_0 is present,

$$b(2\rho|\mu) = -\frac{1}{2} \sum_{j=1}^{\ell} d_j + \frac{k}{2}\frac{\ell}{2} - \sum_{j=1}^{\ell} (j-1)d_j + \frac{k}{2}\binom{\ell}{2}.$$

We conclude from comparing these with expressions occurring in $\overline{bC}$ that if μ is any weight other than λ of the irreducible representation of $g(V)$ of highest weight λ, and if v_μ is a weight-vector of weight μ, so that $v_\mu\varphi(e_j \otimes e_{-j}) = d_j v_\mu$ for all j, then

$$\text{(26)} \qquad b[(\lambda + 2\rho|\lambda) - (\mu + 2\rho|\mu)]$$

$$= R(v_\lambda) - R(v_\mu),$$

where

$$v_\lambda C^{(0)} = R(v_\lambda)v_\lambda, v_\mu C^{(0)} = R(v_\mu)v_\mu,$$

and where, specifically,

$$\text{(27)} \qquad R(v_\mu) = \frac{1}{2} \sum_{j=1}^{\ell} d_j^2 - \frac{k}{2} \sum_{j=1}^{\ell} d_j - \sum_{j=1}^{\ell} jd_j, \quad \text{if } \varepsilon = -1;$$

$$= \frac{1}{2} \sum_{j=1}^{\ell} d_j^2 - \frac{k}{2} \sum_{j=1}^{\ell} d_j - \sum_{j=1}^{\ell} (j-1)d_j, \quad \text{if } \varepsilon = 1, \text{ even case};$$

$$= \frac{1}{2} \sum_{j=1}^{\ell} d_j^2 - \frac{k+1}{2} \sum_{j=1}^{\ell} d_j - \sum_{j=1}^{\ell} (j-1)d_j, \quad \text{if } \varepsilon = 1, \text{ odd case}.$$

($R(v_\lambda)$ is obtained by replacing d's by c's throughout.)

In particular, if $\mu \neq \lambda$ we conclude that

$$\text{(28)} \qquad R(v_\lambda) > R(v_\mu).$$

Now we pass to $A_k(V_\infty)$. Let $\Lambda = (b_1, b_2, \ldots)$ determine a maximal right ideal b_Λ, where here we assume $s(\Lambda) = k$. By b) and d) of §3, this has the effect of assuring that for all sufficiently large j, $\varphi(e_j \otimes e_{-j})$ annihilates v_Λ, the image of $1 \in A_k$ in $A_k/b_\Lambda = M_\Lambda$. Thus for any finite sequence $X_{-\alpha_{i_1}}, \ldots, X_{-\alpha_{i_t}}$, we have that for sufficiently large j, $\varphi(e_j \otimes e_{-j})$ centralizes all these $X_{-\alpha_{i_s}}$ and annihilates v_Λ. In particular, <u>every weight vector in M_Λ is annihilated by almost all</u> $\varphi(e_j \otimes e_{-j})$. We know that v_Λ is annihilated by all $\varphi(e_i \otimes e_j)$, where $i+j > 0$.

Next let $i + j > 0$, $i \le j$. If j is greater than the absolute value of any index $1,\ldots,|i_t| + 1$ involved in $X_{-\alpha_{i_1}},\ldots,X_{-\alpha_{i_t}}$, then it follows from our ordering of weights that

(28) $v_\Lambda X_{-\alpha_{i_1}},\ldots,X_{-\alpha_{i_t}} \varphi(e_i \otimes e_j)$ is either zero or belongs to a weight higher than that of v_Λ, and we have seen that is impossible. Hence j is bounded above if (28) is to be non-zero. But then our weight vector is annihilated by all but finitely many terms in the "renormalized Casimir operator"

$$C' = \frac{1}{2}\sum_{j=1}^{\infty} \varphi(e_j \otimes e_{-j})^2 - \frac{k}{2}\sum_{j=1}^{\infty} \varphi(e_j \otimes e_{-j}) - \sum_{j=1}^{\infty} j\varphi(e_j \otimes e_{-j})$$

$$- \frac{1}{2}\sum_{j=1}^{\infty} (e_j \otimes e_j)\varphi(e_{-j} \otimes e_{-j}) - \sum_{0<i<j} \varphi(e_i \otimes e_j)\,(e_{-j} \otimes e_{-i})$$

$$+ \frac{1}{2}\sum_{0<i<j} \varphi(e_{-i} \otimes e_j)\varphi(e_{-j} \otimes e_i), \text{ if } \varepsilon = -1,$$

and likewise in the cases where $\varepsilon = 1$, replacing the upper index of summation "ℓ" by "∞" in (11) and (12).

By appeal to the finite-dimensional case, or by direct calculation, one sees that C', regarded as operator on M_Λ, maps v_Λ to $R(v_\Lambda)v_\Lambda$, where $R(v_\Lambda)$ is defined as was $R(v_\lambda)$, now with sums like $\frac{1}{2}\sum_{j=1}^{\infty} c_j^2$ involved. Because $k = s(\Lambda)$, all but finitely many c_j are zero, and these sums are finite. Moreover, C' commutes with the action (on a given weight vector) of all $X_{-\alpha_i}$. Thus we have the

<u>Proposition 3:</u> <u>For every</u> $v \in M_\Lambda$, vC' <u>is defined and is equal to</u> $R(v_\Lambda)v$.

In particular, if v_μ is a weight vector belonging to a weight $(d_0,d_1,\ldots)$, then almost all $d_j = 0$, and $v_\mu C^{(0)} = R(v_\mu)v_\mu$, where $R(v_\mu)$ is defined as was $R(v_\mu)$, but without the upper limit "ℓ" on the sums. Thus we have

$$R(v_\Lambda)v_\mu = v_\mu C' = v_\mu C^{(0)} + \sum_{0<i<j} v_\mu\varphi(e_{-i} \otimes e_j)\varphi(e_{-j} \otimes e_i)$$

$$+ \varepsilon \sum_{0<i<j} v_\mu\varphi(e_i \otimes e_j)\varphi(e_{-j} \otimes e_{-i}) - \frac{1}{2}\sum_{j=1}^{\infty} v_\mu\varphi(e_j \otimes e_j)\varphi(e_{-j} \otimes e_{-j}),$$

where the understanding is that the last term is present only when $\varepsilon = -1$, and $i = 0$ is admissible only when $\varepsilon = 1$ and we are in the odd case. Because

$v_\mu C^{(0)} = R(v_\mu)v_\mu$ and from (26)-(28), we have

Lemma 1: Let $k = s(\Lambda)$ as above, and let v_μ be a weight vector of M_Λ belonging to a weight strictly less than that of v_Λ. Then v_μ is a positive rational multiple of

$$\sum_{0<i<j} v_\mu \varphi(e_{-i} \otimes e_j)\varphi(e_{-j} \otimes e_i) + \varepsilon \sum_{0<i<j} v_\mu \varphi(e_i \otimes e_j)\varphi(e_{-j} \otimes e_{-i})$$

$$- \frac{1}{2} \sum_{j=1}^{\infty} v_\mu \varphi(e_j \otimes e_j)\varphi(e_{-j} \otimes e_{-j}),$$

where the sum is finite.

5. Hermitian Contravariant Forms.

In the context of §2, let F be the complex field $\mathbb{C}$. We fix an integer k for which $A_k = A_k(V_\infty)$ is non-zero, and we let $\bar{A}_k$ be an algebra which has the same additive and multiplicative structure (on the same set) as A_k, but with multiplication by complex scalars defined by $\lambda \mathbin{\bar{\cdot}} a = \bar{\lambda}a$ (scalar multiplication by $\bar{\lambda}$, acting in A_k). Then $\bar{A}_k$ is again a $\mathbb{C}$-algebra with unit.

Let $\mathfrak{f}$ be an additive automorphism of V_∞ stabilizing each V_r, and such that $\mathfrak{f}(\lambda v) = \bar{\lambda}\mathfrak{f}(v)$ for all $\lambda \in \mathbb{C}$, $v \in V_\infty$; $(\mathfrak{f}(u),\mathfrak{f}(v)) = \overline{(u,v)}$ for all u,v; and $\mathfrak{f}(\mathfrak{f}(v)) = v$ for all v. Then there is an additive map $\gamma: V_\infty \otimes_{\mathbb{C}} V_\infty \to \bar{A}_k$ such that

$$\gamma(u \otimes v) = \varphi(\mathfrak{f}(u) \otimes \mathfrak{f}(v)).$$

For $\lambda \in \mathbb{C}$,

$$\gamma(\lambda(u \otimes v)) = \varphi(\mathfrak{f}(\lambda u) \otimes \mathfrak{f}(v)) = \bar{\lambda}\varphi(\mathfrak{f}(u) \otimes \mathfrak{f}(v))$$

$$= \lambda \mathbin{\bar{\cdot}} \gamma(u \otimes v),$$

so that γ is $\mathbb{C}$-linear. Now γ satisfies identities *i)*-*iv)* of §1 (recall that $\lambda^* = \lambda$ here). This is perfectly straightforward; for *iv)* one shows by induction that, with γ_t as usual,

$$\gamma_t(u_1,\ldots,u_t;v_1,\ldots,v_t) = \varphi_t(\mathfrak{f}(u_1),\ldots,\mathfrak{f}(u_t);\mathfrak{f}(v_1),\ldots,\mathfrak{f}(v_t)),$$

using simply that $\mathfrak{f}([u,v,w]) = [\mathfrak{f}(u),\mathfrak{f}(v),\mathfrak{f}(w)]$.

It follows that there is an algebra-homomorphism $\theta: A_k \to \bar{A}_k$ mapping $\varphi(u \otimes v)$ to $\gamma(u \otimes v) = \varphi(\mathfrak{f}(u) \otimes \mathfrak{f}(v))$, i.e., a conjugate-linear ring-endomorphism of A_k with this effect. Then θ^2 is an algebra-endomorphism of A_k, mapping $\varphi(u \otimes v)$ to $\theta(\varphi(\mathfrak{f}(u) \otimes \mathfrak{f}(v))) = \varphi(\mathfrak{f}^2(u) \otimes \mathfrak{f}^2(v)) = \varphi(u \otimes v)$, so θ^2 is the identity.

Now let η be the involutorial anti-automorphism of A_k of §1. Then the composites $\theta\circ\eta$ and $\eta\circ\theta$ are both conjugate-linear anti-automorphisms of A_k, and each sends $\varphi(u\otimes v)$ to $\varepsilon\varphi(\mathfrak{f}(v)\otimes\mathfrak{f}(u))$. Thus θ and η commute on A_k, and <u>the composite</u> $\sigma = \theta\circ\eta = \eta\circ\theta$ <u>is a conjugate-linear ring anti-automorphism of</u> A_k <u>satisfying:</u> σ^2 = identity; $\sigma(\varphi(u\otimes v)) = \varepsilon\varphi(\mathfrak{f}(v)\otimes\mathfrak{f}(u))$, for all $u,v\in V_\infty$.

The particular $\mathfrak{f}$ that we shall choose will be the unique conjugate-linear map $\mathfrak{f}:V_\infty\to V_\infty$ such that $\mathfrak{f}(e_i) = e_{-i}$ for all $i\geq 0$, $\mathfrak{f}(e_{-i}) = \varepsilon e_i$ for all $i\geq 0$. This clearly has the properties desired, and we have $\sigma(\varphi(e_i\otimes e_j)) = \varepsilon\varphi(e_{-j}\otimes e_{-i})$ for all i,j, where σ is the composite involution above, whenever both $i,j\geq 0$ or both $i,j<0$, and in all cases if $\varepsilon = 1$. If $\varepsilon = -1$, and if i,j have opposite signs,

$$\sigma(\varphi(e_i\otimes e_j)) = -\varepsilon\varphi(e_{-j}\otimes e_{-i}) = \varphi(e_{-j}\otimes e_{-i}).$$

Let $M = M_\Lambda \cong A_k(V_\infty)/b_\Lambda$ be a right A_k-module, irreducible as in our earlier considerations. Then $\mathrm{Hom}_{\mathbb{C}}(M_\Lambda,\mathbb{C})$ is a vector space with definition $\langle m|\lambda\cdot m'\rangle = \bar\lambda\langle m|m'\rangle = \langle\bar\lambda m|m'\rangle$ for all $m\in M_\Lambda$, $m'\in\mathrm{Hom}_{\mathbb{C}}(M_\Lambda,\mathbb{C})$. If $a\in\bar A_k^{op}$, define $m'a$ by $\langle m|m'a\rangle = \langle ma|m'\rangle$. Then $m'(a\cdot b)$ satisfies
$\langle m|m'(a\cdot b)\rangle = \langle m(a\cdot b)|m'\rangle = \langle mba|m'\rangle = \langle mb|m'\cdot a\rangle = \langle m|(m'\cdot a)\cdot b\rangle$, and
$\langle m|m'(\lambda\cdot a)\rangle = \langle m|m'(\bar\lambda a)\rangle = \langle m(\bar\lambda a)|m'\rangle = \langle(\bar\lambda m)a|m'\rangle = \langle\bar\lambda m|m'\cdot a\rangle = \bar\lambda\langle m|m'\cdot a\rangle = \langle m|\lambda\cdot(m'\cdot a)\rangle = \langle m|(\lambda\cdot m')\cdot a\rangle$.

Thus $\mathrm{Hom}_{\mathbb{C}}(M_\Lambda,\mathbb{C})$ is a right $\bar A_k^{op}$-module, and hence a right A_k-module, with $m'*a = m'\cdot\sigma(a)$. In particular, if i,j have the same sign, or in general if $\varepsilon = 1$, $m'*\varphi(e_i\otimes e_j) = \varepsilon m'\varphi(e_{-j}\otimes e_{-i})$, while if i,j have opposite signs and $\varepsilon = -1$, $m'*\varphi(e_i\otimes e_j) = m'\varphi(e_{-j}\otimes e_{-i})$. Thus, except when $\varepsilon = -1$ and i,j have the same sign, $\langle m|m'*\varphi(e_i\otimes e_j)\rangle = \langle m|m'\varphi(e_{-j}\otimes e_{-i})\rangle = \langle m\varphi(e_{-j}\otimes e_{-i})|m'\rangle$, and in general

$$\langle m|m'*a\rangle = \langle m|m'\sigma(a)\rangle = \langle m\sigma(a)|m'\rangle$$

for all $m\in M_\Lambda$, $m'\in\mathrm{Hom}_{\mathbb{C}}(M_\Lambda,\mathbb{C})$, $a\in A_k$.

For each weight (c) of M, let $N_{(c)}$ be the subspace of $\mathrm{Hom}_{\mathbb{C}}(M,\mathbb{C})$ whose kernels contain $\sum_{(d)\neq(c)} M_{(d)}$. $N_{(c)}$ thus identifies with the dual space $M^*_{(c)}$, and is finite-dimensional. Let

$$M^{(0)} = \sum_{(c)\text{ a weight of } M} N_{(c)},$$

a subspace of $\mathrm{Hom}_{\mathbb{C}}(M,\mathbb{C})$ which is the direct sum of the $N_{(c)}$.

We note that $M^{(0)}$ <u>is a right</u> A_k<u>-submodule of</u> $\mathrm{Hom}_{\mathbb{C}}(M,\mathbb{C})$ <u>in our action</u>, because each $N_{(c)} * A_k \subset M^{(0)}$. This is seen from

$$N_{(c)} * \varphi(e_{-(j+1)} \otimes e_j) \subset N_{(c)-(0,\ldots,0,\overset{j}{1},\overset{j+1}{-1},0,\ldots)},$$

$$N_{(c)} * \varphi(e_{-j} \otimes e_{j+1}) \subseteq N_{(c)-(0,\ldots,-1,1,0\ldots)},\ j = 1,2,\ldots$$

$$N_{(c)} * \varphi(e_{-1} \otimes e_{-1}) \subset N_{(c)+(0,2,0\ldots)},\ N_{(c)} * \varphi(e_1 \otimes e_1) \subset N_{(c)-(0,2,0\ldots)}$$

if $\varepsilon = -1$. (Recall that the notation is: $(c) = (c_0, c_1, \ldots)$, even when c_0 is superfluous.);

$$N_{(c)} * \varphi(e_{-2} \otimes e_{-1}) \subset N_{(c)+(0,1,1,0\ldots)};$$

$$N_{(c)} * \varphi(e_2 \otimes e_1) \subset N_{(c)-(0,1,1,0\ldots)} \quad \text{if } \varepsilon = 1, \text{ even case;}$$

$$N_{(c)} * \varphi(e_0 \otimes e_1) \subset N_{(c)-(0,1,0\ldots)} \quad \text{and}$$

$$N_{(c)} * \varphi(e_0 \otimes e_{-1}) \subset N_{(c)+(0,1,0\ldots)} \quad \text{if } \varepsilon = 1, \text{ odd case;}$$

and $N_{(c)} * \varphi(e_j \otimes e_{-j}) \subset N_{(c)}$ in all cases. Thus $M^{(0)}$ is stable under the action of a generating set for A_k, and is therefore an A_k-submodule.

Next note that $M^{(0)}$ is a sum of weight-spaces, with $N_{(c)}$ being the subspace of weight (c) in $M^{(0)}$. By virtue of the above, it suffices to show that if $m \in M_{(c)}$, $n \in N_{(c)}$, then for all $j > 0$, $<m|n * \varphi(e_j \otimes e_{-j})> = c_j<m|n>$. By definition $<m|n * \varphi(e_j \otimes e_{-j})> = <m|n\sigma(\varphi(e_j \otimes e_{-j}))> = <m\sigma(\varphi(e_j \otimes e_{-j}))|n> = <m\varphi(e_j \otimes e_{-j})|n> = c_j<m|n>$, regardless of sign of ε.

Now $M^{(0)}$ <u>is an irreducible right A_k-module.</u> For any submodule R must also be a sum of its weight spaces $R_{(c)} = N_{(c)} \cap R$. If $R \neq M^{(0)}$, then $R_{(c)} \neq N_{(c)}$ for some (c), and by duality of $N_{(c)}$ and $M_{(c)}$ (finite-dimensional spaces) there is $0 \neq m \in M_{(c)}$ with $<m|R_{(c)}> = 0$. For $(d) \neq (c)$ we clearly have $<m|N_{(d)}> = 0$, so $<m|R> = 0$ by the last paragraph. But then $R^{\perp} = \{x \in M|<x|R> = 0\}$ is a nonzero submodule of M, hence is M, and this is only possible if $R = 0$.

Next note that if $\hat{1} \in M^{(0)}$ satisfies $<\bar{1}|\hat{1}> = 1$, $\hat{1} \in N_{(c)}$, where (c) is the weight of $\bar{1}$ (the canonical image in $M = A_k/b_\Lambda$ of $1 \in A_k$), then $\hat{1}$ <u>generates the right A_k-module $M^{(0)}$</u> <u>and is annihilated by</u> b_Λ.

Clearly there is $\hat{1}$ as specified and $\hat{1}$ generates $M^{(0)}$ by the irreducibility of $M^{(0)}$. Moreover $\hat{1}$ belongs to the weight (c), and this determines that $\hat{1}$ is annihilated by the generators of b_Λ of the form $H_j - b_j 1$, $j = 1,2,\ldots$. If $i,j > 0$, $\hat{1} * \varphi(e_i \otimes e_j)$ belongs to a "weight" $(c) - (0,\ldots,0,1,0,\ldots,0,1,0,\ldots)$, $(c) - (0,\ldots,0,2,0,\ldots)$ if $i = j$, and this is greater than (c) in our ordering so cannot be a weight. A similar argument shows that $\hat{1} * \varphi(e_i \otimes e_j) = 0$ whenever $i + j > 0$, so that $\hat{1}$ is annihilated by a set of generators for b_Λ.

We have now proved

<u>Proposition 4:</u> <u>There is an isomorphism $\omega: M \to M^{(0)}$ of right A_k-modules under which $\bar{1}$ is sent to $\hat{1}$, and every isomorphism $M \to M^{(0)}$ is a scalar multiple of ω. We have $\omega(M_{(c)}) = N_{(c)}$ for each weight (c).</u>

The existence of the isomorphism and the last assertion are clear from the above. The remaining statement follows from the fact that the subspace of M annihilated by all $\varphi(e_i \otimes e_j)$, $i+j > 0$, is $\mathbb{C}\bar{1}$, and therefore that the corresponding space for $M^{(0)}$ is $\mathbb{C}\hat{1}$.

For $m,n \in M$, let $h(m,n) = \langle m|\omega(n)\rangle \in \mathbb{C}$. Then h is biadditive, is linear in the first variable, and

$$h(m,\lambda n) = \langle m|\omega(\lambda n)\rangle = \langle m|\lambda\cdot\omega(n)\rangle = \bar{\lambda}\langle m|\omega(n)\rangle$$

$$= \bar{\lambda}h(m,n), \quad \text{for } \lambda \in \mathbb{C}.$$

That is, h is sesquilinear. Moreover, $h(ma,n) = \langle ma|\omega(n)\rangle = \langle m|\omega(n)*\sigma(a)\rangle = \langle m|\omega(n\sigma(a))\rangle = h(m,n\sigma(a))$.

<u>Proposition 5:</u> <u>The sesquilinear form h is positive-definite hermitian on M. Weight-spaces belonging to distinct weights are h-orthogonal.</u>

<u>Proof</u>: To see that h is hermitian, fix $n \in M$, and let ψ_n be the mapping $M \to \mathbb{C}$ mapping $m \in M$ to $\overline{h(n,m)} = \overline{\langle n|\omega(m)\rangle}$. As in the proof of sesquilinearity of h, one sees that $\psi_n \in \mathrm{Hom}_{\mathbb{C}}(M,\mathbb{C})$. Now if $n \in M_{(d)}$ for some weight (d), the fact that ω maps weight-spaces to spaces of the same weight shows that $h(n,M_{(d')}) = 0$ for $(d') \neq (d)$, so that $\psi_n \in M^{(0)}$. Moreover, $n \to \psi_n$ is linear: $M \to \mathrm{Hom}_{\mathbb{C}}(M, \mathbb{C})$ (with our conjugate-action of scalars on $\mathrm{Hom}_{\mathbb{C}}(M, \mathbb{C})$), so maps M linearly into $M^{(0)}$. If $a \in A_k$, we have that $\psi_n * a = \psi_n\cdot\sigma(a)$ sends $m \in M$ to $\langle m|\psi_n \sigma(a))\rangle$

$$= \overline{\langle m\sigma(a)|\psi_n\rangle} = \overline{\langle n,\omega(m\sigma(a))\rangle}$$

$$= \overline{\langle n|\omega(m) * \sigma(a)\rangle} = \overline{\langle n|\omega(m)\cdot a\rangle}$$

$$= \overline{\langle na|\omega(m)\rangle} = \langle m|\psi_{na}\rangle .$$

Thus $n \longmapsto \psi_n$ is a non-trivial homomorphism of A_k-modules $M \to M^{(0)}$, so is an isomorphism, by irreducibility. Thus there is $\tau \in \mathbb{C}$, $\tau \neq 0$, with $\psi_n = \tau\omega(n)$ for all $n \in M$. That is, $\overline{\langle n|\omega(m)\rangle} = \langle m|\tau\cdot\omega(n)\rangle = \bar{\tau}h(m,n)$ for all m,n, or $\overline{h(n,m)} = \bar{\tau}h(m,n)$. With $m = n = \bar{1}$, we have $h(\bar{1},\bar{1}) = \langle\bar{1}|\hat{1}\rangle = 1$, so $\bar{\tau} = 1$, and $\tau = 1$. Thus $\psi_n = \omega(n)$, and h <u>is hermitian</u>. The assertion about orthogonality of weight-spaces follows from properties of ω.

Finally, to see that h is positive definite, we must show its restriction to each weight space is positive definite. Our linear ordering on weights is discrete, and there are only finitely many weights greater than a specified one. This follows from the fact that if $c_\nu = 0$ for all $\nu > j$, then any nonzero element $v_\Lambda X_{-\alpha_{i_1}} \ldots X_{-\alpha_{i_t}}$ in which some $i_s > j$ belongs to a lower weight than (c), and then from the fact that the $g(V_r)$-module generated by v_Λ is finite-dimensional for each r. The unique highest weight is that of $\bar{1}$, and

$$h(\lambda\bar{1},\lambda\bar{1}) = \lambda\bar{\lambda} > 0 \quad \text{if} \quad \lambda \neq 0.$$

Thus the restriction of h to the highest weight space is positive definite. Now let (c) be a weight other than the highest one, and assume inductively we have proved that the restriction of h to each weight space of higher weight than (c) is positive definite. Let $0 \neq v \in M_{(c)}$; then for some pair of indices i,j, with $i+j > 0$, $v\varphi(e_i \otimes e_j) \neq 0$.

By Lemma 1, there is a positive rational γ such that

$$\begin{aligned}\gamma v = &\sum_{0<i<j} v\varphi(e_{-i} \otimes e_j)\varphi(e_{-j} \otimes e_i) \\ &+ \varepsilon \sum_{0\le i\le j} v\varphi(e_i \otimes e_j)\varphi(e_{-j} \otimes e_{-i}) \\ &- \frac{1}{2} \sum_{j>0} v\varphi(e_j \otimes e_j)\varphi(e_{-j} \otimes e_{-j}),\end{aligned}$$

the sum being finite. Thus

$$\begin{aligned}\gamma h(v,v) = h(\gamma v,v) = &\sum_{0<i<j} h(v\varphi(e_{-i} \otimes e_j)\varphi(e_{-j} \otimes e_i),v) \\ &+ \varepsilon \sum_{0\le i<j} h(v\varphi(e_i \otimes e_j)\varphi(e_{-j} \otimes e_{-i}),v) \\ &- \frac{1}{2} \sum_{j>0} h(v\varphi(e_j \otimes e_j)\varphi(e_{-j} \otimes e_{-j}),v) \\ = &\sum_{0<i<j} h(v\varphi(e_{-i} \otimes e_j),v\sigma(\varphi(e_{-j} \otimes e_i))) \\ &+ \varepsilon \sum_{0\le i<j} h(v\varphi(e_i \otimes e_j),v\sigma(\varphi(e_{-j} \otimes e_{-i}))) \\ &- \frac{1}{2} \sum_{j>0} h(v\varphi(e_j \otimes e_j),v\sigma(\varphi(e_{-j} \otimes e_{-j}))).\end{aligned}$$

In each sum the term $v\varphi(e_{-i} \otimes e_j)$, $v\varphi(e_i \otimes e_j)$, $v\varphi(e_j \otimes e_j)$ belongs to a weight greater than (c), and at least one of these is different from zero.

In the first sum, $\sigma(\varphi(e_{-j} \otimes e_i)) = \varepsilon\varphi(\mathfrak{z}(e_i) \otimes \mathfrak{z}(e_{-j})) = \varepsilon\varphi(e_{-i} \otimes \varepsilon e_j)$ $= \varphi(e_{-i} \otimes e_j)$. In the second sum $\sigma(\varphi(e_{-j} \otimes e_{-i})) = \varepsilon\varphi(e_i \otimes e_j)$; in the third, $\varepsilon = -1$ and $\sigma(\varphi(e_{-j} \otimes e_{-j})) = \varepsilon\varphi(e_j \otimes e_j) = -\varphi(e_j \otimes e_j)$. Thus $\gamma h(v,v)$ is a sum of terms $h(w,w)$, where each w is a weight vector belonging to a weight strictly greater than (c) with at least one $w \neq 0$. It follows that $h(v,v) > 0$, and our proof by induction is complete.

It will be observed that the involution σ stabilizes each $\mathfrak{g}(V_r)$. The elements x of $\mathfrak{g}(V_\infty)$ for which $\sigma(x) = -x$ act as skew-hermitian transformations of $\mathcal{M}$, while those with $\sigma(y) = y$ act as hermitian transformations. For each such $z = x$ or y, and for each $m \in \mathcal{M}$, z stabilizes a finite-dimensional subspace containing m, as does a $\mathfrak{g}(V_r)$ containing z. Thus

$$m \exp(z) \text{ makes sense, and } \exp(x)$$

is a unitary automorphism of $\mathcal{M}$, $\exp(y)$ a positive-definite hermitian transformation of $\mathcal{M}$. There would seem to be promising analogies with the finite-dimensional case to be investigated in these actions on the pre-Hilbert space $\mathcal{M}$, and possibly with regard to its completion.

Bibliography

[1] Bourbaki, N., Groupes et algèbres de Lie, Ch. 7-8. Hermann, Paris, 1975.

[2] Humphreys, J., Introduction to Lie algebras and representation theory. Springer Verlag, Berlin-Heidelberg-New York, 1972.

[3] Jacobson, N., Lie Algebras. Interscience, New York, 1962. Reprinted Dover, New York, 1979.

[4] Seligman, G., Generalized even Clifford algebras. Jour. of Algebra 82 (1983), 398-458.

[5] ________, Higher even Clifford algebras. Sém. d'Algèbre P. Dubreil et M-P. Malliavin. Proceedings, Paris, 1982. Springer Lecture Notes, Vol. 1029, pp. 159-191. Springer-Verlag, Berlin-Heidelberg-New York, 1983.

G. B. Seligman
Department of Mathematics
Yale University
Box 2155 Yale Station
New Haven, Connecticut 06520
USA

LA RESULTANTE DE DEUX POLYNOMES

Alain LASCOUX

Résumé - Après avoir rétabli l'équation résultante de deux polynômes dans son sexe originel, nous montrons que la théorie des fonctions symétriques fournit sans calcul de nombreuses expressions déterminantales de la dite résultante, et plus généralement, du plus grand commun diviseur.

La résultante de deux polynômes a été l'objet de maintes recherches depuis le 18ème siècle, étant au coeur de la théorie de l'élimination. On la trouvait développée dans tous les livres d'algèbre du siècle dernier. De nos jours, sa part s'est malheureusement réduite à une expression déterminantale attribuée à Sylvester. Voilà sans doute qui explique la redécouverte périodique de certaines de ses propriétés [cf. Gerber, Householder]. Nous puiserons la majorité de nos références dans les cinq livres de [Muir] qui résument la majorité des formules déterminantales de la littérature (mathématique) jusqu'en 1920.

1. SYMETRIE -

La définition de la résultante de deux polynômes remonte à Euler et Bezout, si ce n'est plus tôt. Supposant les polynômes P et Q totalement factorisés, i.e. $P = \Pi_{a \in A}(x-a)$ et $Q = \Pi_{b \in B}(x-b)$, A et B étant les deux ensembles de racines, alors par définition

$$(1,1) \qquad R(A,B) = \Pi_{a \in A, b \in B}(a-b)$$

est la résultante et tout polynôme en les $a \in A$ et les $b \in B$ qui s'annule quand P et Q ont une racine commune est divisible par R(A,B).

Bien entendu, on ne va pas chercher les racines des deux polynômes - problème dont [Ruffini], puis Galois ont montré la difficulté algébrique ; c'est la symétrie en les $a \in A$, ainsi qu'en les $b \in B$ qui fournit la méthode la plus directe d'évaluer la résultante. Nous avons donc besoin de quelques définitions relatives aux fonctions symétriques [Macdonald].

Soit M une matrice, k un entier positif, $I = (i_1,\ldots,i_k) \in \mathbb{Z}^k$, $J \in \mathbb{Z}^k$. On note $M_{I,J}$ le mineur pris sur les lignes $i_1+1,\ldots,i_k+k$ et les colonnes $j_1+1,\ldots,j_k+k$ (on pose $M_{I,J} = 0$ si l'une de ces lignes ou colonnes n'existe pas ; pour $I = (0,\ldots,0) = 0^k = J$, on a le mineur "initial" pris sur les k premières lignes et k premières colonnes).

A toute série formelle $\Sigma_{j \geqslant 0}\, z^j\, S_j$, on associe la <u>matrice de ses coefficients</u>

$$[S] = [S_{j-i}]_{i,j \geqslant 0}$$

avec $S_j = 0$ si $j < 0$, et les <u>fonctions de Schur</u>

$$(1.2) \qquad S_{J/I} = [S]_{I,J}\,, \quad I,J \in \mathbb{Z}^k.$$

Dans le cas où $I = 0^k$, on écrit S_J.

Il faut noter que le produit des séries formelles ou l'inversion correspondent au produit des matrices, ou à l'inversion. La formule de Binet-Cauchy (cf. note) pour les mineurs d'un produit de matrices, ou la formule de Jacobi pour les mineurs de la matrice inverse donnent donc les fonctions de Schur pour le produit de deux séries ou l'inverse d'une série [Littlewood].

Etant donnés deux ensembles A,B d'indéterminées (nous dirons deux "alphabets"), les <u>fonctions de Schur</u> $S_{J/I}(A-B)$ sont les fonctions de Schur de la série :

$$(1.3) \qquad \Pi_{b\in B}(1-zb)/\Pi_{a\in A}(1-za) = \Sigma\, z^j\, S_j(A-B)$$

Avec ces notations, les polynômes $\Pi(x-a)$ et $\Pi(x-b)$ s'écrivent respectivement $S_m(x-A)$ et $S_n(x-B)$, m et n étant les cardinaux de A et B.

On note A+B l'union <u>disjointe</u> de deux alphabets. Les fonctions symétriques sont écrites en supposant toutes les indéterminées différentes, quitte à spécialiser ensuite.

La <u>formule de Cauchy</u>, dont on a fait le deuxième axiome des λ-anneaux, est l'expression de $\Pi(a-b)$ (ou, de manière équivalente, de

$\Pi(1-ab)$, de $1/\Pi(1-ab)$) comme fonction symétrique en les $a \in A$, et les $b \in B$:

(1.4) $\Pi_{a\in A, b\in B}\ (a-b) = S_{n^m}(A-B)$ (cf. note).

Factorisant la série $\Sigma\, S_j(A-B)$ en $\Sigma S_j(A).\Sigma S_k(-B)$, on transforme comme suit la formule précédente :

(1.5) $\Pi(a-b) = \Sigma\, S_{n^m/I}(A)\ .\ S_I(-B)$,

somme sur toutes les partitions $I = 0 \leq i_1 \leq i_2 \leq \ldots \leq i_m$.

Brioschi (1854 ;[Muir] II p.349) a donné l'expression (1.4) du résultant en définissant les $S_j(A-B)$ par :

(1.6) $S_{j+m-n}(A-B) = \Sigma_{a\in A}\ a^j\ Q(a)/P'(a)$.

2. METHODE D'EULER ET BEZOUT -

On doit aux mathématiciens sus-nommés un principe très puissant pour fournir des multiples du résultant : construire, pour un entier r quelconque, r polynômes de degré $\leq$ r-1 qui s'annulent simultanément pour toute racine commune à $S_m(x-A)$ et $S_n(x-B)$. Alors le déterminant des coefficients de ces polynômes est un multiple du résultant, puisqu'il s'annule en même temps que lui (la matrice des coefficients admet le vecteur propre $1\ C\ldots C^{r-1}$ pour toute "racine commune" $C \in \{A \cap B\}$). On veille usuellement à ce que les r polynômes soient linéairement indépendants et tels que le déterminant de leurs coefficients soit de degré minimum (= mn) en A,B.

Soit par commodité m=n ; Bezout note que pour $0 \leq j \leq n-1$, les n "équations dérivées"

(2.1) $P_j = S_j(x-A).S_n(x-B) - S_j(x-B).S_n(x-A)$

sont des polynômes dont le degré ne dépasse pas n-1 et qui s'annulent pour toute racine commune dans $\{A \cap B\}$. La matrice des coefficients des P_j a été appelée "Bezoutien" par Sylvester ; (voir aussi [Jacobi],[Muir] I p.485 et [Muir] III p.343).

Le Bezoutien n'est pas identique à la matrice

$[S_{n+j-i}(A-B)]_{1 \leq i,j \leq n}$

il s'obtient à partir de cette dernière par multiplication à gauche et à droite par la matrice triangulaire $[S_{j-i}(-A)]_{1 \leq i,j \leq n}$, le déterminant du Bezoutien est donc bien $R(A,B)$. Par exemple, pour $m = n = 3$, $R(A,B) = S_{333}(A-B)$ tandis que le Bezoutien est égal à

$$\begin{vmatrix} 1 & -A & S_2(-A) \\ 0 & 1 & -A \\ 0 & 0 & 1 \end{vmatrix} \cdot \begin{vmatrix} S_3(A-B) & S_4(A-B) & S_5(A-B) \\ S_2(A-B) & S_3(A-B) & S_4(A-B) \\ S_1(A-B) & S_2(A-B) & S_3(A-B) \end{vmatrix} \cdot \begin{vmatrix} 1 & -A & S_2(-A) \\ 0 & 1 & -A \\ 0 & 0 & 1 \end{vmatrix}$$

$$= \begin{vmatrix} S_3(-B)-S_3(-A) & BS_3(-A)-AS_3(-B) & S_2(-A)S_3(-B)-S_2(-B)S_3(-A) \\ S_2(-B)-S_2(-A) & S_3(-B)-S_3(-A)+BS_2(-A)-AS_2(-B) & BS_3(-A)-AS_3(-B) \\ A-B & S_2(-B)-S_2(-A) & S_3(-B)-S_3(-A) \end{vmatrix}$$

Sylvester préfère éliminer $1, x, \ldots, x^{m+n-1}$ entre les polynômes $S_m(x-A)$, $S_{m+1}(x-A), \ldots, S_{m+n-1}(x-A)$; $S_n(x-B), \ldots, S_{m+n-1}(x-B)$ et obtient ainsi le déterminant d'ordre $m+n$ suivant :

$$(2.2) \qquad \left| \begin{matrix} S_0(-A) & \ldots & S_{m+n-1}(-A) \\ \vdots & & \vdots \\ S_{-n+1}(-A) & \ldots & S_m(-A) \\ S_0(-B) & \ldots & S_{m+n-1}(-B) \\ \vdots & & \vdots \\ S_{-m+1}(-B) & \ldots & S_n(-B) \end{matrix} \right| \begin{matrix} \left.\vphantom{\begin{matrix}a\\a\\a\end{matrix}}\right\} n \\ \left.\vphantom{\begin{matrix}a\\a\\a\end{matrix}}\right\} m \end{matrix}$$

(voir Jacobi (1835), [Muir] I p.214 ; Sylvester (1840), [Muir] I p.236 ; Richelot (1840), [Muir] I p.238 ; Cauchy (1840), [Muir] I p.242).

Il faut noter que, $S_{m+j}(x-A)$ étant égal à $x^j\, S_m(x-A)$ pour tout $j \geq 0$, ce déterminant s'obtient en prenant n lignes formées par les coefficients de $S_m(x-A)$, en décalant d'un cran, d'une ligne à l'autre, puis m lignes formées par les coefficients de $S_n(x-B)$.

Le développement du déterminant de Sylvester suivant ses n premières lignes donne une somme de produits de fonctions de Schur de $-A$ et de $-B$; si l'on revient à A, ce qui correspond au changement d'indice indiqué en note, on retrouvera très exactement l'expression (1.5) de la résultante. Par exemple, la résultante

de $S_3(x-A)$ et $S_2(x-B)$ est égale à $S_{222}(A-B) =$
$S_{222}(A).S_{000}(-B) + S_{122}(A).S_{001}(-B) + S_{112}(A).S_{011}(-B)$
$+ S_{022}(A).S_{002}(-B) + S_{012}(A).S_{012}(-B) + S_{111}(A).S_{111}(-B)$
$+ S_{002}(A).S_{022}(-B) + S_{011}(A).S_{112}(-B) + S_{001}(A).S_{122}(-B)$
$+ S_{000}(A).S_{222}(-B)$

$$= \begin{vmatrix} 1 & -A & S_2(-A) & S_3(-A) & 0 \\ 0 & 1 & -A & S_2(-A) & S_3(-A) \\ 1 & -B & S_2(-B) & 0 & 0 \\ 0 & 1 & -B & S_2(-B) & 0 \\ 0 & 0 & 1 & -B & S_2(-B) \end{vmatrix}$$

3. LA RESULTANTE PAR DIVISION -

Le reste de la division d'un polynôme par un autre est divisible par le p.g.c.d. de ces deux polynômes. Au lieu de prendre, comme précédemment dans la méthode de Sylvester, les polynômes $S_m(x-A)$, $S_{m+1}(x-A)$, $S_{m+2}(x-A), \ldots$, on peut se contenter de leurs restes par $S_n(x-B)$.

Voyons tout d'abord comment les fonctions de Schur interviennent dans la division.

Lemme 3.1. - Soient p un entier dans $\mathbb{Z}$, A et B deux alphabets, n le cardinal de B. Alors
$S_p(x-A) - S_{p-n}(x-A+B)\, S_n(x-B) = \Sigma_0^{n-1} S_j(x-B).S_{p-j}(B-A)$.

En effet, $x-A = (x-B) + (B-A)$, et donc

$$S_p(x-A) = \sum_0^{\infty} S_j(x-B) \,.\, S_{p-j}(B-A) = \sum_0^{n-1} + \sum_n^{\infty} .$$

Or, puisque B est de cardinal n,

$$h \geqslant 0 \Longrightarrow S_{n+h}(x-B) = x^h S_n(x-B) ;$$

la sous somme $\sum_n^{\infty} S_j(x-B).S_{p-j}(B-A)$ est donc égale à

$$S_n(x-B)\sum_0^\infty x^h S_{p-n-h}(B-A) = S_n(x-B)\, S_{p-n}(x+B-A) \quad \square$$

Le membre de droite de 3.1 est donc le <u>reste</u> de la division de $S_p(x-A)$ par le polynôme $S_n(x-B)$.

En particulier, la matrice des coefficients des restes des polynômes $S_m(x-A), S_{m+1}(x-A),\ldots,S_{m+n-1}(x-A)$ dans la base $S_0(x-B),\ldots,S_{n-1}(x-B)$ est la matrice $S_{m^n}(B-A)$ déjà vue en 1.4.

<u>Corollaire 3.2</u> - <u>Pour tout</u> $p \geq 0$, $x^p \equiv x^{n-1} S_k(B) - x^{n-2} S_{1k}(B) + x^{n-3} S_{11k}(B) - x^{n-4} S_{111k}(B) + \ldots$ <u>modulo le polynôme</u> $S_n(x-B)$, <u>avec</u> $k = p-n+1$.

Le corollaire résulte de (3.1) pour un alphabet A vide ; on vérifie que la somme $\sum_0^{n-1} S_j(x-B).S_{p-j}(B) = \sum_{i,j} x^{n-1-i} S_{i+j+1-n}(-B)\, S_{p-j}(B)$ se transforme en ce qu'exige (3.2) moyennant l'identité (cf. note)

$$S_i(-B)\, S_k(B) + S_{i-1}(-B)\, S_{k+1}(B) + \ldots + S_0(-B)\, S_{k+i}(B) = (-1)^i S_{1^i k}(B) \quad \square$$

<u>Remarque</u> : Si $0 \leq p \leq n-1$, un seul terme dans la somme (3.2) est non nul : c'est $(-1)^{n-1-p} x^p S_{1^{n-1-p}k}(B) = x^p S_{0^{n-p}}(B) = x^p$, comme il se doit.

Les restes d'autres ensembles de n polynômes peuvent donner des fonctions intéressantes. C'est le cas des $S_{1^j}(B-A-x).S_m(x-A)$, $0 \leq j \leq n-1$, comme le montre le lemme suivant (qui est équivalent, par linéarité, à 3.1).

<u>Lemme 3.3</u> - <u>Soient</u> A <u>et</u> B <u>deux alphabets de cardinaux respectifs</u> m,n, <u>et</u> j <u>un entier</u> ≥ 0. Alors $S_{1^j}(B-A-x).S_m(x-A) \equiv S_{1^j m}(B-A) + (x-B)\, S_{1^j m-1}(B-A) + S_2(x-B)\, S_{1^j m-2}(B-A) + \ldots + S_{n-1}(x-B)\, S_{1^j m-n}(B-1)$ <u>modulo</u> $S_n(x-B)$.

Le déterminant des coefficients $S_{1^j m-i}(B-A)$, $0 \leq i,j \leq n-1$ des n-1 premiers restes est en fait égal à $(-1)^{\binom{n}{2}} S_{m^n}(B-A)$; plus généralement, toute fonction de Schur $S_{J/I}(B-A)$ peut s'exprimer comme un déterminant en les fonctions "équerres" $S_{1\ldots1k}(B-A)$ [L & P].

On peut répéter la division et diviser $S_n(x-B)$ par le reste de la division de $S_m(x-A)$ par $S_n(x-B)$, et ainsi de suite ; les coefficients des restes successifs sont, comme plus haut, des fonctions de Schur (les "équerres" 1...1 k sont à remplacer par les partitions plus générales 1...1 k...k, r fois k pour le r-ième reste). Cela donne la résultante comme un déterminant, de la taille que l'on désire, en fonctions de Schur. C'est à [Wronski] qu'il faut attribuer les relations entre fonctions de Schur nécessaires à la récurrence sur r.

4. RACINES COMMUNES -

Le résultant $R(A,B)$ est nul si le p.g.c.d de $S_m(x-A)$ et $S_n(x-B)$ est de degré ≥ 1 ; chacune des matrices que nous avons données dans les paragraphes précédents est alors de rang non maximal. On peut faire mieux que de constater la nullité du déterminant des dites matrices. Par exemple, le p.g.c.d s'exprime à l'aide des mineurs d'ordre approprié d'une quelconque des dites matrices. Les calculs reposent essentiellement sur le lemme élémentaire suivant.

Tout d'abord, étant donné un entier n, $J \in \mathbb{Z}^n$ et n couples d'alphabets $A_1,B_1,\ldots,A_n,B_n$, on note $S_J(A_1-B_1,\ldots,A_n-B_n)$ la fonction (que l'on dira encore de Schur)

$$|S_{j_k+k-h}(A_k-B_k)|_{1 \leq h,k \leq n}$$

<u>Lemme 4.1</u>. <u>Soient</u> $C_1,\ldots,C_n$ <u>des alphabets tels que</u> $\mathrm{card}(C_1)\leq n-1$, $\mathrm{card}(C_2) \leq n-2,\ldots$ $\mathrm{card}(C_n) \leq 0$. <u>Alors</u>

$$S_J(A_1-B_1,\ldots,A_n-B_n) = |S_{j_k+k-h}(A_k-B_k-C_h)|$$

i.e. <u>on peut soustraire</u> C_1 <u>à la première ligne</u>, C_2 <u>à la deuxième</u>, ... <u>sans changer la valeur de la fonction de Schur</u>.

En effet, pour tout entier p et tout alphabet C de cardinal r

$$S_p(A-B-C) = \sum_{j=0}^{r} S_{p-j}(A-B)\; S_j(-C).$$

La première ligne du déterminant S_J est donc modifiée par adjonction des suivantes puisque $\mathrm{card}(C_1) \leq n-1$; il en est de même des autres, la valeur du déterminant est inchangée.□

Le lemme 4.1 permet de factoriser certaines fonctions de Schur ; cette factorisation est le point central de nombreux articles en théorie de l'élimination ; cependant, il ne semble pas que l'énoncé général en ait été breveté avant Berele et Regev, et c'est donc à ces auteurs , selon les recommandations de [Grothendieck] que nous attribuerons le lemme suivant ([Pomey] donne par exemple le cas $I = 0\ldots0$, $J = 0\ldots01..1$; [Giambelli (1908)] y ajoute le cas $J = 0..0\,k$).

<u>Lemme 4.2</u> : <u>Soient</u> p <u>un entier positif</u>, A,B <u>deux alphabets de cardinaux respectifs</u> m,n, $I \in \mathbb{N}^p$, $J \in \mathbb{N}^m$. <u>Alors</u>

$$S_{i_1\ldots i_p\; n+j_1\ldots n+j_m}(A-B) = S_I(-B).S_J(A).R(A,B).$$

Notons en passant que le lemme 4.2 donne des invariants de la suite récurrente $\{S_j(A-B)\}$ comme déterminants isobares; c'est une autre approche de cette factorisation dont Euler connaissait des cas particuliers.

Soient A',B',C trois alphabets disjoints, de cardinaux respectifs -r,n-r,r, $A = A'+C$, $B = B'+C$. L'ensemble des "racines communes" de $_m(x-A)$ et $S_n(x-B)$ est $C = A \cap B$, i.e. le p.g.c.d des deux olynômes est $S_r(x-C)$.

Ecrivant $C = A-A'$, on peut développer le p.g.c.d :

$$S_r(x-C) = \Sigma_j\, S_{r-j}(x-A).S_j(A').$$

es $S_j(A')$ nous sont données par le lemme 4.2, au facteur $R(A',B')$ rès. En effet

$$\begin{aligned} S_{\square+\,o\ldots oj}(A-B) &= S_{\square+o\ldots oj}(A'-B') \\ &= S_j(A').R(A',B') = S_j(A').S_\square(A'-B') \\ &= S_j(A').S_\square(A-B) \end{aligned}$$

n notant □ et □ + o...oj les partitions

$$\underbrace{-r,\ldots,n-r}_{m-r \text{ fois}} \quad \text{et} \quad \underbrace{n-r,\ldots,n-r}_{m-r-1 \text{ fois}},\ n-r+j.$$

Toujours d'après 4.2,

$$o \leq p < r \Rightarrow S_{(n-p)^{m-p}}(A-B) = S_{(n-p)^{r-p}}(-B').R(A',B').S_{(n-p)^{m-r}}(A')$$

est nul puisque $S_{n-p}(-B') = 0$, n-p étant strictement supérieur au cardinal de B' ; comme réciproquement, pour un sous-alphabet de B, B' de cardinal $\leq n-r \Leftrightarrow S_n(-B) = S_{n-1}(-B) = \ldots = S_{n-r+1}(-B)=0$, on obtient finalement le lemme suivant.

<u>Lemme 4.3</u> : <u>Pour que le p.g.c.d des polynômes</u> $S_m(x-A)$ <u>et</u> $S_n(x-B)$ <u>soit de degré r, il faut et il suffit que les déterminants</u> $S_{n^m}(A-B),\ldots,S_{(n-r+1)^{m-r+1}}(A-B)$ <u>soient nuls et que</u> $S_\square(A-B) \neq 0$, <u>avec</u> $\square = (n-r)^{m-r}$.

<u>Dans ce cas, le p.g.c.d est, au facteur</u> $S_\square(A-B)$ <u>près, égal à</u>

$$\sum_o^r S_{r-j}(x-A) \,.\, S_{\square + ooj}(A-B).$$

La fonction $S_{\square + ooj}(A-B)$ est le mineur d'indice $(r^{m-r},o..oj)$ de la matrice $[S_{n-h+k}(A-B)]_{1 \leq h,k \leq m}$, c'est-à-dire est un mineur de la première matrice qui nous a, en 1.4, donné la résultante ; on n'éprouve pas plus de difficulté à identifier $S_{\square + ooj}(A-B)$ à un mineur des autres matrices dont le déterminant est la résultante. Prenons par exemple le déterminant de Sylvester (2.2) :

<u>Lemme 4.4</u> : <u>Soit</u> r <u>le degré du p.g.c.d de</u> $S_m(x-A)$ <u>et</u> $S_n(x-B)$. <u>Alors ce dit p.g.c.d est égal, à une constante près, au déterminant d'ordre</u> m+n-2r

$$\begin{vmatrix} S_o(-A) & \ldots & S_{m+n-2-2r}(-A) & S_{m+n-1-r}(x-A) \\ \vdots & & \vdots & \vdots \\ S_{-n+r+1}(-A) & \ldots & S_{m-r-1}(-A) & S_m(x-A) \\ S_o(-B) & \ldots & S_{m+n-2-2r}(-B) & S_{m+n-1-r}(x-B) \\ \vdots & & \vdots & \vdots \\ S_{-m+r+1}(-B) & \ldots & S_{n-r-1}(-B) & S_n(x-B) \end{vmatrix}$$

<u>Démonstration</u> : on peut développer suivant les puissances de x et essayer de reconnaître les fonctions déjà rencontrées en (4.3). Il

est plus simple de constater que ce déterminant s'annule pour toute racine x commune aux deux polynômes de départ, puisqu'alors $A-x$ est un alphabet de card $m-1$ d'où $S_m(x-A) = S_{m+1}(x-A) = \ldots = 0$ et similairement $S_n(x-B) = S_{n+1}(x-B) = \ldots = 0$, i.e. la dernière colonne est nulle. En outre, le déterminant est un polynôme de degré $\leqslant r$, les puissances plus élevées de x s'éliminant en soustrayant à la dernière colonne x^o fois la première $+ x^1$ fois la seconde .... . Enfin, on trouve comme coefficient de x^r la fonction non nulle $S_{(n-r)^{m-r}}(A-B)$ $(= R(A',B'))$, en développant le déterminant suivant ses $n-r$ premières lignes comme on l'a fait, pour $r=0$, en 2.2. □

Il y a de nombreuses manières de présenter les calculs de ce paragraphe ; bien des travaux du siècle dernier en théorie de l'élimination s'y rapportent. Le point de vue le plus instructif est sans doute celui d'Euler et Bezout, à savoir l'étude pour tout entier $p \geqslant 0$ de l'espace vectoriel V_p engendré par les polynômes $S_m(x-A), S_{m+1}(x-A), \ldots, S_{m+p}(x-A)$; $S_n(x-B), \ldots, S_{m+p}(x-B)$ (en supposant $m \geqslant n$) ; cela revient à filtrer l'idéal engendré par $S_m(x-A)$ et $S_n(x-B)$; si $n-p > r$ = degré du p.g.c.d, alors $V_p = \mathbb{C}[x] f_p(x)$, où $f_p(x)$ est le p+1-ième reste de la division de $S_m(x-A)$ par $S_n(x-B)$; f_{n-r-1} est le p.g.c.d et $V_{n-r-1} = V_{n-r} = \ldots =$ Idéal engendré par $S_m(x-A)$ et $S_n(x-B)$.

Un historique tant soit peu complet est hors de portée de cet exposé ; voir l'index du cinquième volume de [Muir] et les pages 321-338 de [Pascal].

Voici toutefois quelques jalons :
[Jacobi] donne les restes et quotients successifs, ce qui revient à réduire la fraction rationnelle $S_m(x-A) / S_n(x-B)$ en fraction continue. La méthode de [Wronski] pour la séparation des racines d'un polynôme selon que leur module est supérieur à 1 ou non conduit aux mêmes calculs. L'expression des restes comme des déterminants du type 4.4 se trouve dans Sylvester (1840,[Muir] I p.238 ; [Muir] II p.340 ; oeuvres, I p.58 ; I p.429-586) , Faa de Bruno (1859,[Muir] III p.327). Abel et Liouville ont donné le p.g.c.d lorsqu'il est de degré 1 ; [Brioschi] obtient le p.g.c.d dans le cas général à l'aide des dérivées de R(A,B) en tant que fonction des coéfficients des deux polynômes. [Trudi] (1862, [Muir] III p.340 donne le lemme 4.4. [Pomey] (1888) exprime le dit p.g.c.d à l'aide successivement des $S_j(A-B)$,

du Bezoutien et de la matrice de Sylvester ; dans [Weber] (p.188 et s) sont utilisées les équerres $S_{1^j k}(A-B)$. [Giambelli (1908)] montre que l'on peut aussi employer un déterminant en les $S_i(A)$ et $S_j(B)$ (au lieu des $S_i(-A)$ et $S_j(-B)$ dans le déterminant de Sylvester).

5. AUTRES POINTS DE VUE -

Cayley (1853, Muir II p.138) donne une interprétation intéressante du Bezoutien : il considère la forme quadratique

$$Q(x,y) = [S_m(x-A)\, S_n(y-B) - S_m(y-A)\, S_n(x-B)] \,/\, (x-y) =$$

$$[1\ x \ldots x^{m-1}]\ [R_{ij}]_{1\leq i,j\leq m}\ [1\ y \ldots y^{m-1}]^{tr}$$

supposant $m \geqslant n$. Cette forme quadratique s'annule quelque soit y si x est une racine commune des deux polynômes ; le déterminant de Q est donc nul quand la résultante l'est. En fait, $(x-y)\, Q(x,y)$ est égale au déterminant du produit de $\begin{bmatrix} 1 & x & \ldots & x^m \\ 1 & y & \ldots & y^m \end{bmatrix}$ par la transposée $\begin{bmatrix} S_o(-A) & \ldots & S_m(-A) \\ S_o(-B) & \ldots & S_m(-B) \end{bmatrix}$; la formule de Binet-Cauchy fournit alors que la matrice $[Q_{ij}]$ est le Bezoutien (voir aussi [Schendel].

Plus généralement, étant donnés r polynômes $P_1,\ldots,P_r$, on étudie la fonction symétrique en $x_1,\ldots,x_r$:

$$|P_i(x_j)| \,/\, \Pi_{i<j}\,(x_i - x_j) \;;$$

voir en particulier l'application à la théorie des polynômes orthogonaux en [L & Shi]

La résultante est un <u>déterminant de Hankel</u> (i.e. une fonction de Schur "rectangle" S_{m^n}). Les identités vérifiées par ces déterminants servent en théorie des polynômes orthogonaux et de l'approximation de Padé [Brezinski].

Une autre approche consiste à considérer le Bezoutien, en fixant un des deux polynômes, comme un morphisme de l'espace des polynômes dans celui des matrices [Orzech].

Plus adapté aux méthodes informatiques est le calcul du Bezoutien de deux polynômes de degré n à l'aide de la valeur en $n+1$ points de ces deux polynômes (Borchardt 1859, [Muir] II p. 147).

Le même Borchardt (1878, [Muir] III p. 161) donne aussi une expression non déterminantale de la résultante. Pour plus de deux polynômes, voir [Giambelli 1908].

6. APPROCHE GEOMETRIQUE -

On peut interpréter en termes géométriques la division d'un polynôme par un autre. En effet, soit B un fibré vectoriel de rang n sur une base quelconque Z, $\mathbb{P}(B) \to Z$ le projectif associé. L'anneau de Grothendieck de $\mathbb{P}(B)$ est le quotient de l'anneau de polynôme $K(Z)[x,x^{-1}]$ par l'idéal engendré par $S_n(x-B)$, x étant la classe du fibré vectoriel "tautologique" de rang 1, et $K(Z)$ l'anneau de Grothendieck de la base. On dispose d'un morphisme de $K(Z)$-modules, (dit morphisme de Gysin) :

$$\Pi_\star : K(\mathbb{P}(B)) \to K(Z).$$

Le reste d'un polynôme $Q(x)$ quelconque s'écrit, à l'aide de ce morphisme :

$$Q(x) \equiv \sum_{o}^{n-1} x^{n-1-k}\, \Pi_\star(Q(x) . S_k(x-B) . x^{1-n})$$

(cf. [L & S1] pour la généralisation à la variété de drapeaux). Le calcul de la résultante consiste à trouver un polynôme $Q(x)$ de degré $\leq n-1$ (unique à constante près) tel que $Q(x).S_m(x-A)$ soit congru à une constante différente de 0 modulo $S_n(x-B)$ (A est considéré comme la classe d'un fibré de rang m). On a déjà vu au paragraphe 4 que :

$$Q(x) = \begin{vmatrix} S_o(-A) & \dots & S_{m+n-2}(-A) & x^{n-1} \\ \vdots & & \vdots & \vdots \\ S_{1-n}(-A) & \dots & S_{m-1}(-A) & x^o \\ S_o(-B) & \dots & S_{m+n-2}(-B) & 0 \\ \vdots & & \vdots & \vdots \\ S_{1-m}(-B) & & S_{n-1}(-B) & 0 \end{vmatrix}$$

et l'on peut retrouver, à l'aide du morphisme $\Pi_\star$ que
$Q(x).S_m(x-A) \equiv R(A,B)$ modulo $S_n(x-B)$.

Pour respecter la symétrie en A et B, on peut faire appel à la grassmannienne $G_n(\mathbb{C}^{m+n}) = G$ que l'on plonge diagonalement dans $G \times G$: $\Delta \hookrightarrow G \times G$; sur G on dispose d'un module noyau tautologique de rang m, disons E, et d'un quotient F de rang n ; les classes de Chern de fibrés vectoriels dans l'anneau de cohomologie de $G \times G$ peuvent être considérés comme des polynômes : on pose $c(p_1^\star(E)) = \Pi_{a \in A}(1+a)$ et $c(p_2^\star(F)) = \Pi_{b \in B}(1+b)$, p_1 et p_2 étant les deux projections de $G \times G$ sur G. Alors la classe de la diagonale Δ dans l'anneau de cohomologie est la classe de Chern maximale $C_{mn}(p_1^\star(E) \otimes p_2^\star(F^v)) = \Pi(a-b) = R(A,B)$. Toutes les formules des paragraphes précédents se traduisent en termes géométriques ; par exemple, le développement (1.5) exprime que la classe de Δ dans la cohomologie de $G \times G$ est égale à la somme de produits de cycles de Schubert d'indices complémentaires, les fonctions de Schur s'interprétant maintenant comme les cycles fondamentaux de la grassmannienne (cycles de Schubert) ; pour des groupes plus généraux que $Gl(\mathbb{C}^n)$, ceci est le théorème de Chevalley.

On peut choisir comme [Giambelli] un point de départ plus algébrique : celui des idéaux de mineurs de matrices. On retrouve résultante, fonctions de Schur et grassmannienne !

7. UNE GENERALISATION -

Tout ce qui précède est l'étude de la fonction $R(A,B)=\Pi(a_i-b_j)$ pour deux ensembles $A = \{a_1,\ldots,a_m\}$ et $B = \{b_1,\ldots,b_n\}$. Supposons $m = n$. La moitié des termes dans la factorisation de la résultante :

$$R_{1/2}(A,B) = \Pi_{i+j \leq n} (a_i - b_j)$$

conduit aux mêmes types de propriétés que la résultante. Ainsi $R_{1/2}(A,B)$ se développe en une somme de produits de polynômes de Schubert [L & S 1], ce qui permet une interpolation de Newton à plusieurs variables [L & S 2]. L'interprétation géométrique fait appel aux variétés de drapeaux [L-S] ; d'autres approches sont possibles, par exemple les tableaux (de Young) ou la classification des décompositions réduites dans le groupe symétrique. Ce qui sous-entend tous ces points de vue, c'est l'action du groupe symétrique sur l'anneau des polynômes, l'algèbre libre, la variété de drapeaux.

NOTE -

La théorie des fonctions symétriques a été pendant longtemps essentiellement rattachée à celle des déterminants isobares ; l'interprétation en terme de caractères de groupe linéaire et symétrique n'est venue qu'avec les travaux de Frobenius, Young et Schur. Cesdites fonctions symétriques imposent un décalage dans les indices par rapport aux conventions prédominantes : si M est une matrice, et $I,J \in \mathbb{Z}^n$, alors $M_{I,J}$ est le mineur pris sur les lignes $i_1+1,\ldots,i_n+n$ et colonnes $j_1,\ldots,j_n+n$, ainsi que nous l'avons dit au premier paragraphe. A permutation de lignes (si toutes différentes) près, on peut supposer que I est une partition, i.e. une suite d'entiers croissante au sens large : $0 \leq i_1 \leq i_2 \leq \ldots \leq i_n$. Le produit de matrices se définit par

$$[MN]_{i,j} = \Sigma_h M_{i,h} . N_{h,j}$$

la sommation est sur $h \in \mathbb{N}$, une fois étendue par des 0 les matrices en cause). Les mineurs de MN s'écrivent tout aussi simplement, à l'aide de la formule de Binet-Cauchy :

$$\forall n, \forall I,J \in \mathbb{Z}^n , [MN]_{I,J} = \Sigma_H M_{I,H} . N_{H,J}$$

somme sur toutes les partitions H (bien entendu, il y a seulement un nombre fini de termes non nuls dans la somme si les matrices sont finies).

Si M est une matrice carrée (de déterminant égal à 1 pour simplifier), et si N est l'image de M^{-1} par rapport à l'anti-diagonale 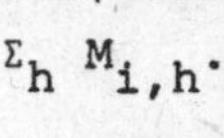 , alors pour tout couple de partitions $I,J \in \mathbb{N}^n$,

$$M_{I,J} = (-1)^{|J|-|I|} N_{I^\sim,J^\sim}$$

c'est la formule de Jacobi.

($I^\sim$ désigne la partition dite conjugée ou transposée de I, i.e. celle obtenue en transposant la représentation planaire de I, et $|I| = i_1 + i_2 + \ldots$ est le <u>poids</u> de I, cf [Macdonald] p.2).

Enfin, le développement d'un déterminant suivant un sous-ensemble de lignes a été obtenu par Laplace ; supposons la matrice M coupée en deux parties N,P 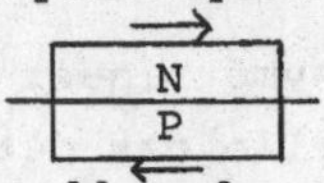 et les colonnes de N numérotées de gauche à droite, celles de P de droite à gauche.
Alors

$$\det(M) = \Sigma\ (-1)^{|J|}\ N_{o\ldots o,J} \cdot P_{o\ldots o,J^{\sim}}$$

somme sur toutes les partitions J (la notation $N_{o\ldots o,J}$ signifie que l'on prend le mineur sur <u>toutes</u> les lignes de N, et colonnes J).

Revenons aux fonctions de Schur. D'après la définition (1.3), la matrice $[S_{j-i}(A-B)]$ est le produit des matrices $[S_{j-i}(A)]$ et $[S_{j-i}(-B)]$, en d'autres termes,

$$S_j(A-B) = \Sigma\ S_{j-h}(A) \cdot S_h(-B).$$

On en déduit, par la formule de Binet-Cauchy,

$$S_{J/I}(A-B) = \Sigma\ S_{H/I}(A) \cdot S_{J/H}(-B)$$

somme sur toutes les partitions H.

Posant $B = A$ dans (1.3), on voit que $\Sigma\ S_j(A) \cdot \Sigma\ S_i(-A) = 1$, c'est-à-dire que les matrices $[S_{j-i}(A)]$ et $[S_{j-i}(-A)]$ sont inverses. D'après Jacobi on a donc, pour toute paire I,J de partitions,

$$S_{J/I}(A) = S_{J^{\sim}/I^{\sim}}(-A)\ ;$$

en particulier, pour tout entier k positif,

$$S_{1^k}(A) = (-1)^k\ S_k(-A)$$

La formule de Laplace permet enfin de développer, par lignes ou par colonnes, une fonction de Schur comme une somme de produits de fonctions de Schur. Par exemple, le développement suivant la dernière colonne s'écrit : pour tout $I \in \mathbb{Z}^n$, $j \in \mathbb{Z}$,

$$S_{Ij} = \Sigma_k (-1)^k \, S_{I/1^k} \, . \, S_{j+k}$$

(Ij désigne l'élément $(i_1,\ldots,i_n,j)$ de $\mathbb{Z}^{n+1}$).

En particulier, pour une <u>partition équerre</u> $(n \geqslant 0 \, , \, j \geqslant 1)$,

$$S_{1^n j} = \Sigma_k (-1)^k \, S_{1^{n-k}} \, . \, S_{j+k}$$

identité que l'on peut aussi écrire : pour tout alphabet A,

$$(-1)^n \, S_{1^n j}(A) = \Sigma_k \, S_{n-k}(-A) \, . \, S_{j+k}(A) .$$

BIBLIOGRAPHIE

A. BERELE, A. REGEV - Hook Young diagrams, prepub.

M. BEZOUT - Théorie générale des Equations algébriques, Paris 1779.

C. BREZINSKI - Padé-type Approximants, Birk haûser 1980.

F. BRIOSCHI - Théorie des déterminants, Paris 1856.

Faa de BRUNO - Théorie de l'élimination, Thèse Paris 1856.

H. GERBER - Am.M.M. 91 (1984) 644-646.

Z. GIAMBELLI - Atti Torino 38 (1903) 551-572.

Z. GIAMBELLI - Atti Torino 40 (1904) 1041-1062.

Z. GIAMBELLI - Mem. R. Ist. Lomb. 20 (1904) 101-135.

Z. GIAMBELLI - Rend. Palermo 25 (1908) 131-144.

A. GROTHENDIECK - Récoltes et Semailles, IIème partie, Montpellier 1985.

A.S. HOUSEHOLDER - The numerical treatment of a single non linear equation, Mc Graw Hill 1970.

C.G. JACOBI - Crelle J.15 (1836) 101-124.

[L - S] A. LASCOUX - C.R. Acad. Sc. Paris 295 (1982) 393.

[L & P] A. LASCOUX & P. PRAGACZ - C.R. Acad. Sc. Paris 299 (1984) 955.

[L & Shi] A. LASCOUX & SHI He - C.R. Acad. Sc. Paris 300 (1985) 681.

[L & S1] A. LASCOUX & M.P. SCHÜTZENBERGER in Invariant Theory, Springer L.N. 996.

[L & S2] A. LASCOUX & M.P. SCHÜTZENBERGER in Séminaire d'algèbre M.P. Malliavin 1984-1985, Springer L.N. 1146.

H. LAURENT - Nouv. Ann. M. (3)7 (1888) 60-65.

D.E. LITTLEWOOD - The Theory of group characters, Oxford 1950.

I.G. MACDONALD - Symmetric functions and Hall polynomials, Oxford M. Mono 1979.

T. MUIR - History of determinants, 5 volumes, Dover reprints 1960.

G. ORZECH - Lin. Mult. Alg. 16 (1984) 275-282.

R. RUFFINI - Opere matematiche, reed. Palermo 1915.

E. PASCAL - I determinanti, Hoepli 1923.

E. POMEY - Nouv. Ann. M(3)7 (1888) 66-90 and 407-427.

L. SCHENDEL - Zeit. 38 (1893) 84-94.

J.J. SYLVESTER - Collected Work, Chelsea reprint.

N. TRUDI - Teoria de determinanti, Napoli 1862.

H. WEBER - Traité d'Algèbre supérieure, Paris 1989.

H. WRONSKI - Philosophie de la Technie Algorithmique : Loi suprême et universelle ; réforme des mathématique. Paris 1815-1817.

ON THE CONSTRUCTION OF RESOLUTIONS OF DETERMINANTAL IDEALS : A SURVEY

Piotr Pragacz and Jerzy Weyman

Dedicated respectfully to David A. Buchsbaum

Contents.

Introduction.

The problem of finding the syzygies of an ideal or a module appeared in algebraic geometry during the nineteenth century. For example Study in his book [5] gave the problem of finding the syzygies of Plücker ideals.

Then Hilbert in his famous paper [H] proved the finiteness of the chains of syzygies over the polynomial rings, and also constructed the syzygies for the ideal generated by indeterminates in polynomial ring i.e. the Koszul complex in a explicit manner(*) . He also proved a remarkable theorem which states that any ideal of homological dimension 1 is determinantal i.e. it is the ideal of $n \times n$ minors of $n \times (n+1)$ matrix up to multiplication by a scalar.

The determinantal varieties were studied by Macaulay who used the syzygies that he could obtain to compute their Hilbert functions (and thus their degree and dimension).

Independently, Italian geometers were studying the same varieties (see references in [Se] , for example). It was Gambelli who began a systematic treatement of determinantal varieties from the "cohomological" point of view .

(*) As J.P. Jouanolou has pointed out to us, what is today called the Koszul complex was constructed for the first time rigorously by A. Schönflies in [Sch].

At the end of the fifties various mathematicians came back to this problem i.e. finding the syzygies of $p \times p$ minors of the matrix $X = (X_{ij})$ $1 \leq i \leq m$, $1 \leq j \leq n$ over the ring $R = \mathbb{Z}[X_{ij}]$ where X_{ij} are indeterminates.

The problem was solved by Eagon-Northcott [E-N] in the case of maximal minors. Then the $n-1 \times n-1$ minors of $n \times n$ matrix [G-N] and $n x(n+1)$ matrix [Po] were also resolved. The criteria for exactness of a complex ([P-S],[B-E]) provided good tools for proving the exactness of these complexes. However, the main problem - finding the syzygies of $p \times p$ minors in the general case was still unsolved.

In 1977 Lascoux in his thesis $[L]_1$ solved this problem in general in characteristic 0 case i.e. he resolved the $p \times p$ minors over $R = K[X_{ij}]$ where K is a field of characteristic 0. His solution used two new key ideas. One was the use of the representations of the general linear group. In this way the fact that the ideal of $p \times p$ minors is invariant under the linear transformations of the rows and columns of a matrix was exploited. The other idea was to use derived functors given by a classical geometrical construction to generalize Koszul complex. This was done first by Kempf in [K].

This survey is an account of the recent development of the theory of syzygies initiated in Lascoux's thesis. In particular the resolutions of other classical ideals are treated (although the Plücker ideals are still difficult to tackle). We mention also the characteristic free results that are known. One should note that the technique reviewed in this paper can be applied to the structure theory of resolutions (see forthcoming paper $[P-W]_2$).

To begin with let R be a polynomial ring over a field or over the integers and let $\mathfrak{a}$ be a homogenous ideal in R. Our problem is to describe the minimal graded R-resolution

$$\mathcal{F}: \ldots \to \mathcal{F}^i \quad \ldots \to \mathcal{F}^1 \to \mathcal{F}^0 = R \quad R/\mathfrak{a} \to 0$$

of $R/\mathfrak{a}$. Notice that in the first case $\mathcal{F}$ always exists and is unique up to an isomorphism. In the second case on can show that if $\mathcal{F}$ exists, then it is unique up to an isomorphism, the problem of existence being open.

The authors thank to J.P. Jouanolou and A. Lascoux of valuable comments concerning historical aspects of the theory of syzygies.

I. USE THE GEOMETRY TO OBTAIN SYZYGIES .

Let R be a polynomial ring over a field and let d be a homogenous ideal in R . Moreover let $X = \operatorname{Spec} R$, $Y = \operatorname{Spec} R/\mathfrak{a} \subset X$ and let R_+ stands for the ideal generated by all indeterminates in R .

The case of the method described in this section can be summarized as follows : add indeterminates to transform the equations of Y , so as to obtain a complete intersection, thus a Koszul complex ; then eliminate the added indeterminates to get the looked for syzygies.

Effective computations need some background of combinatorial nature , which is closely connected to the classical Schur functions.

1. Algebro-geometric background (*) .

To add indeterminates, one needs in fact a geometrical constructions involving schemas which are quasi-projective and not affine in general. The formal procedure can be described as follows.

Let $\bar{X} = X \times G$ be the product variety with its 2 projections

$$X \xleftarrow{\pi} \bar{X} \xrightarrow{\tau} G .$$

Let M be a graded R-module.

Theorem. Assume the existence of an $\mathcal{O}_{\bar{X}}$-module N , of a negatively graded $\mathcal{O}_{\bar{X}}$-resolution $P^\cdot \to N \to 0$ of N , and of an $\mathcal{O}_{\bar{X}}$-resolution $B^{\cdot\cdot}$ of $P^\cdot$ (with $P^\cdot \to B^{\cdot\cdot}$) in the category of complexes.

If the following properties are satisfied :

1. $H^o(\bar{X},N) = N$, $H^k(\bar{X},N) = 0 \quad k > 0$,
2. $P^i = \tau^* Q^i$ for some $\mathcal{O}_G$-module Q^i ($i \leqslant 0$) .
3. $B^{m,\cdot} = \tau^* A^{m,\cdot}$ for some complex of acyclic $\mathcal{O}_G$-modules $A^{m,\cdot}$ ($m \leqslant 0$) .

3'. The horizontal differential in $\pi_* B^{\cdot\cdot} \otimes (R/_{R_+})$ vanisches then

$$\mathcal{F}^i = \bigoplus_{m+k=i} H^k(\bar{X},P^m) .$$

(*) This is a straightening of methods in $[P]_1$, [J-P-W] .

For most of the applications one can take M to be equal to $\mathcal{O}_Y$. The module N plays the role of the structure sheaf of a certain locally complete intersection Z, which lifts Y. Usually G stands for a certain Grassmann variety and the Q^i are homogenous vector bundle over G.

For this construction is needed one simple property of spectral sequences of hypercohomology, namely :

Lemma. If assumption 1 is satisfied then the total complex of $\pi_*(B^{\cdot\cdot})$ denoted $\mathrm{tot}\,\pi_*(B^{\cdot\cdot})$ is a resolution of N (usually non-minimal).

Proof of the theorem. It is well known that $\mathcal{F}^i = \mathrm{Tor}^R_{-i}(M, R/R_+) \otimes_{R/R_+} R$. To compute this Tor, let us use the resolution

$$\mathrm{tot}\,\pi_* B^{\cdot\cdot} \longrightarrow M \longrightarrow 0 .$$

We have

$$H_i(\mathrm{tot}\ \pi_* B^{\cdot\cdot} \otimes R/R_+) = \bigoplus_{m+k=i} H^k(\pi_* B^{m,\cdot} \otimes R/R_+) \quad \text{(by 3')}$$

$$= \bigoplus_{m+k=i} H^k(A^{m,\cdot})$$

$$= \bigoplus_{m+k=i} H^k(G, Q^m) \quad \text{(since } A^{m,j} \text{ are acyclic).}$$

Thus $\mathcal{F}^i = \bigoplus_{m+k=i} H^k(G,Q^m) \otimes R = R^k\, \pi_* P^m$, which completes the proof.

Taking the classes in the $K(X)$ gives :

Corollary. $[\mathcal{O}_Y] = \sum_i (-1)^{|i|} [\sum_{m+k=i} R^k \pi_* P^m]$ in $K(X)$.

The cohomology $H^k(G,Q)$, when G is a Grassmann variety and Q stands for a homogenous bundle over it, is described by Bott's theorem. The formulation of this theorem needs the notion of Schur modules. Moreover the construction of $B^{\cdot\cdot}$ in many cases requires Schur complexes too. Therefore we need some informations from

2. Combinatorics.

Recall that Schur functors S_I are indexed by partitions, i.e. by weakly decreasing sequences $I = i_1 \geq i_2 \geq \ldots \geq i_m > 0$ of natural numbers. Each partition I can be represented by a diagram D_I with i_1 boxes in the first row, i_2 boxes in the second row, etc. For a partition I define its length to be $\ell g I = m$, its weight to be $|I| = \Sigma i_k$ and its rank to be $\mathrm{rk} I$ = length of the diagonal in D_I.

Example. $I = 421$, $\ell g(I) = 3$, $|I| = 7$, $\mathrm{rk}\, I = 2$

$D_I =$ 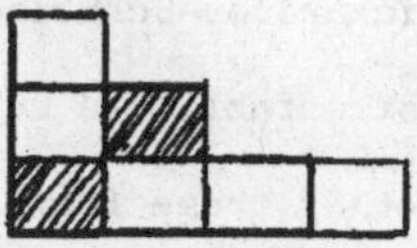

Now let $|I| = n$ and let $e_I \in Q[\Sigma_n]$ be the idempotent corresponding to the irreducible representation of Σ_n indexed by I. The Schur functor S_I is defined by

$$E \longrightarrow S_I(E) = e_I E^{\otimes n}$$

where Σ_n acts on $E^{\otimes n}$ by permuting the factors.
Restricted to the vector space V, $S_I(V)$ gives the polynomial representation of $GL(V)$ which can also be interpreted as $H^0(\mathrm{Flag}\, V, L_1^{i_1} \otimes \ldots \otimes L_r^{i_r})$ where $L_1, \ldots, L_{\dim(V)}$ are the tautological line bundles over the variety Flag (V) of full flags in V.

This definition can be extended to the complexes thanks to the work of Nielsen [N]. Note that the module M itself, respectively the map $M \to N$ of modules can be considered as complexes

$$\underset{}{0} \to \ldots \to 0 \to \underset{0}{M} \to 0 \to \ldots \to 0$$

and respectively

$$0 \to \ldots \to 0 \to \underset{-1}{M} \to \underset{0}{N} \to 0 \to \ldots \to 0 .$$

Now for a given complex $X^\bullet$ we can treat $X^{\bullet\otimes n}$ as an Σ_n-module (taking sign into account

$$x\otimes y \to (-1)^{\deg x\cdot\deg y}\, y\otimes x)$$

and define $S_I(X^\bullet) = e_I X^{\bullet\otimes n}$. For example the exterior power of $\underset{-1}{M}\to\underset{0}{N}$ is given by

$$0\to \underset{-n}{S_n N}\to S_{n-1}M\otimes M\to \dots \to \underset{-i}{S_i M\otimes \Lambda^{n-i}N}\to \dots \to \underset{-1}{M\otimes\Lambda^{n-1}N}\to \underset{0}{\Lambda^n N}\to 0 .$$

There exists a nice combinatorial rule for multiplying arbitrary Schur modules called "Littlewood-Richardson-Schützenberger rule". The special cases $S_I\otimes S_n$ or $S_I\otimes\Lambda^n$ were first established by Pieri in the geometric context. For the detailed description of these formulas, we refer the reader to [L-S]. The same rule holds for Schur complexes.

Now we state Bott's theorem in the version adapted for the needs of this paper (compare $[L]_1$ for the details).

Given an ordered pair or natural numbers (a,b) where $a<b$ define a new par $R(a,b) = (b+1,a-1)$ - <u>rectification</u> of (a,b) if $b+1\geq a-1$. In the opposite case write $R(a,b)=\emptyset$. Now given an arbitrary sequence of natural numbers $A=(a_1,\dots,a_n)$ we write $R(A)$ = partition B if B can be obtained by succesive application of the process of rectification of the pairs of the consecutive elements. Write $R(A)=\emptyset$ if such partition B doesn't exist. Let $m(A)$ be the minimal number of rectifications necessary to go from A to B.

<u>Theorem (Bott)</u>. Let $G = G_r(V)$ be the Grassmannian of r-subspaces in an n-dimensional vector space V, over a field of characteristic 0 let R be the tautological bundle and $Q = V_G/R$ the quotient. Then

1° $H^0(G,S_I Q) = S_I V$, $H^i(G,S_I Q) = 0$ for $i>0$

2° $H^n(G,S_I R\otimes S_J Q) = \begin{cases} 0 & , n\neq m(J,I) \\ S_{R(J,I)}V & , n = m(J,I) \end{cases}$

3° $m(\underbrace{0\dots0}_{n-r},I) = r.\mathrm{rank}\, I$.

3. Examples.

There are a lot of geometric constructions which theoretically could be helpful for the computations of syzygies in the sense of Theorem 1. Some of them cannot be used in practice yet - the main reason is the problem of the connecting homomorphism formulated below.

We give here those constructions for which computation was possible.

A. Determinantal ideals of generic matrix (over a field K of characteristic 0).

The problem of describing syzygies in this case was solved by Lascoux $[L]_1$; the resolution in this case was also constructed in different ways by Nielsen [N] and Roberts $[R]_1$. Lascoux used essentially the following construction : Let U and V be two vectors spaces over K . Let us consider $X = \mathrm{Spec}\, S.(U^* \otimes V)$, $G = (\Lambda^p U^* \otimes \Lambda^p V).S_{.}(U^* \otimes V)$, $G = G_{p-1}(V)$, the variety $Z \subset \bar{X} = X \times G$ being defined by the following correspondance :

$(x,L) \in Z$ iff the map $U^* \to V$ induced by x factors through L .

Of course $\pi(Z) \subset Y$. If x is of rank r , then $(x,L) \in Z$ iff $L = \mathrm{Im}\, x$. Thus one can easily show that $\pi : Z \to Y$ is a birational isomorphism. On the other hand Z can be presented as the schema of zeroes of the composit morphism :

$$U_{\bar{X}} \to V_{\bar{X}} \to \tau^* Q .$$

This information allows us to prove that Z is a smooth locally complete intersection of codimension equal to $\mathrm{rk}\, \mathrm{Hom}(U_{\bar{X}}, \tau^* Q)$, and thus Z is a (rational) desingularisation of Y .

In this case, the resolution $P^{\bullet}$ in question is given by the Koszul resolution $\Lambda^{\bullet}(U_{\bar{X}} \otimes \tau^* Q)$. The columns of $B^{\bullet\bullet}$ are the complexes $A^{-m,\bullet} = \Lambda^m[U_G^* \otimes (V_G \to R^*)]$ pulled back on $\bar{X}$ by τ^* , $m = 0,1,2,\ldots$. The differential between consecutive columns is defined by imitation of the Koszul differential with the help of the map

$$\begin{array}{c} U_G^* \otimes (V_G^* \to R^*) \\ \downarrow \\ \mathcal{O}_G[-1] \end{array}$$

instead of the usual cosection. $\mathcal{F}^{\bullet}$ is computed using Cauchy formula : $\Lambda^m(U_{\overline{X}} \otimes Q_{\overline{X}}^*) = \bigoplus_{|I|=m} S_I U_{\overline{X}} \otimes S_I Q_{\overline{X}}^*$ and Bott's theorem. For details see [L.] .

B. Pfaffian ideals of generic antisymmetric matrix (see [J-P],$[P]_1$, [J-P-M]) .

Let V be a vector space over a field K of characteristic 0 . Let us consider $X = \operatorname{Spec} S.(\Lambda^2 V^*)$, $G = \Lambda^{2p} V^* . S.(\Lambda^2 V^*)$. In analogy with the solution of problem A , the following desingularisation of Y seems to be good candidate : $G = G_{2p-2}(V)$ and $Z \subset \overline{X} = X \times G$ defined by the same correspondence as above. But then Z is a locally complete intersection associated with the section of the bundle N involved in the following short exact sequence on X .

$$0 \to R_{\overline{X}} \otimes Q_{\overline{X}} \to N \to \Lambda^2 Q_{\overline{X}} \to 0 .$$

Therefore the calculation of $H^i(G,\Lambda^j N^*)$ requires not only the knowledge of $H^i(G,\Lambda^j(R^* \otimes Q^* + \Lambda^2 Q^*))$ but also requires the control of the connecting homomorphism induced by this sequence. This seems to be impossible at the present stage of the theory (cf. also problem 2 below).

The following construction allows to avoid this difficulty : $G = G_{n-p-1}(V^*)$, $Z \subset X \times G$ is defined by the following correspondence : $(x,L) \in Z \Leftrightarrow L$ is the isotropic subspace of the alternating form induced by x (compare $[P]_1$ or [J-P-W] chap. 3. for the detailed treatment of this construction).

Let us restrict for a moment our attention to the case $p = 2$, which is also the case of the Plücker ideal defining the embedding $G_2(n) \to \mathbb{P}^N$, $N = \binom{n}{2} - 1$. In this case the resolution $P^{\bullet}$ is given by the Koszul resolution $\Lambda^{\bullet}(\Lambda^2 \tau^* R)$ associated with the cosection $\Lambda^2 \tau^* R \to \overline{X} \times \Lambda^2 V^* \to \mathcal{O}_{\overline{X}}$. Note that we have at least 2 possibilities for the construction of $B^{\bullet\bullet}$.

First we can treat the complexes $A^{-m,\bullet}$, where $A^{-m,\bullet} = \Lambda^m(\Lambda^2(V_G^* \to Q))$ $m = 0,1,\dots$, as columns in the double complex $B^{\bullet\bullet}$.

The differential between consecutive columns is defined by imitation of the Koszul differential with the help of the map $\begin{array}{c}\Lambda^2(V_G^* \to Q)\\ \downarrow \\ \mathcal{O}_G[-2]\end{array}$ instead of the usual cosection.

The second possibility is to build the consecutive columns in $B^{\bullet\bullet}$ as $B^{-m,\bullet} = \tau^* A^{-m,\bullet}$ where $A^{-m,\bullet} = \Lambda^m(\Lambda^2 V_G^* \to R\otimes Q)$. Of course this construction comes from the exact sequence $0 \to \Lambda^2 R \to \Lambda^2 V_G^* \to R\otimes Q \to 0$. Maps between consecutive columns can be defined similarly as above and the π_*-acyclity of chains follows easily from Bott's theorem. So the π_*-acyclic resolution in question is not canonical.

For computation of $\mathfrak{F}^\bullet$ one uses in this case Schur-Litllewood's formula decomposing $\Lambda^\bullet(\tau^*\Lambda^2 R)$ in terms of Schur functors. For the details see $[P]_1$,[J-P-W].

C. Minors of a symmetric matrix (see $[L]_2$,[J-P-W]).

Let now $X = \operatorname{Spec} S.(S_2 V)$, $G = S_{\underbrace{2\ldots 2}_{p}} V \cdot S_\bullet(S_2 V)$. Unfortunately no-construction is known, which leads immediately to the result without difficulties concerning connecting homomorphisms. One can overcome these difficulties using two constructions which "aproximate" the syzygies in question. For the details see [J-P-W]. Observe that as usual computing invariants of the orthogonal group is more difficult than with invariants associated with the symplectic group.

D. Plücker ideals defining the grassmannian $G_r(V)$.

$X = \operatorname{Spec} S.(\Lambda^r V^*)$, $S/G = \bigoplus_i S_{\underbrace{i\ldots i}_{r}}(V^*)$. The following construction which generalises the 4×4 Pfaffian case seems to be the most promising. Let $s = n-r+1$ and let $G = G_s(V^*)$. Define the bundle T on $\overline{X}$ by

$$0 \to T \to \Lambda^r(V_{\overline{X}}^*) \xrightarrow{p} R_{\overline{X}} \otimes \Lambda^{r-1} Q_{\overline{X}} \to 0 .$$

Let us notice that p exists since $\mathrm{rk}\, Q = r-1$. The natural cosection $\varphi : \Lambda^r V_X^* \to \mathcal{O}_X$ induces the cosection $\tilde{\varphi}: T \to \Lambda^r(V_X^*) \to \mathcal{O}_X$. Let Z be the subvariety of zeros of $\tilde{\varphi}$.

For instance the syzygies of $G_3(6)$ (in characteristic 0) computed in this way are given as the total complex of the following double complex (the partion I stands for $S_I(V_X)$)

$$
\begin{array}{ccccccccccccc}
777777 & & & & & & & & & & & & \\
\downarrow & & & & & & & & & & & & \\
766665 & \to & \begin{matrix}666654\\765555\end{matrix} & \to & 665544 & & & & & & & & \\
 & & \downarrow & & \downarrow & & & & & & & & \\
 & & \begin{matrix}744444\\555552\end{matrix} & \to & \begin{matrix}554442\\555333\\644433\end{matrix} & \to & \begin{matrix}544332\\544332\end{matrix} & \to & \begin{matrix}444222\\443331\\533322\end{matrix} & \to & \begin{matrix}522222\\33333\end{matrix} & & \\
 & & & & & & & & \downarrow & & \downarrow & & \\
 & & & & & & & & 332211 & \to & \begin{matrix}321131\\22221\end{matrix} & \to & 21111 \\
 & & & & & & & & & & & & \downarrow \\
 & & & & & & & & & & & & \mathcal{O}_X
\end{array}
$$

The main obstruction to compute the syzygies of Plücker ideals is the problem of plethysm : $\Lambda^i(\Lambda^r) = ?$" (see problem 1. below).

4. Differentials (still characteristic zero case).

Although these methods don't give an explicit description of the differentials, some knowledge about them is obtained. The crucial observation is the fact that the graded components of the differentials are natural. This information implies immediately that there are only 2 types of differentials : linear and of degree $p, 2p, p$ in the case of generic minors, pfaffians and symmetric minors respectively. Moreover the syzygies are consequences of the Koszul relations and Laplace type relations for minors and pfaffians.

5. Open questions.

1. Plethysm.

For further applications of these methods, the following problem appear to be of great importance. Decompose $\Lambda^m(S_I E) = \bigoplus_J S_J E$ into its irreducible components (in characteristic 0).

2. The problem of the connecting homomorphisms.

Consider the grassmannian $G = G_r(V)$ with tautological bundle R, and quotient bundle $Q = V_G/R$. Let $0 \to A \xrightarrow{\eta} T \xrightarrow{\varepsilon} B \to 0$ be an exact sequence of homogenous vector bundles on G such that

$A = \bigoplus_{I,J} m(I,J) S_I R \otimes S_J Q$, $B = \bigoplus_{I',J'} m'(I',J') S_{I'} R \otimes S_{J'} Q$. Describe as a function of $(\eta, \varepsilon, I, J, I', J', m, m')$ the connecting homomorphism $H^n(G,B) \xrightarrow{\delta} H^{n+1}(G,A)$.

Note that since δ is a $GL(V)$-morphism, it is a sum of certain homotheties and the zero map. Of course the $GL(V)$-decomposition of $H^n(G,A)$ and $H^n(G,B)$ is given by Bott's theorem.

II. TRACE AND EVALUATION CONSTRUCTIONS.

Although the method described in section I gives the modules coming in the resolutions of determinantal ideals, and some informations about differentials, it does not provide completely the complex which constitutes a resolution ; this can be done by using quite different technique. It stems from [A-B-W]$_1$, where the resolution of $p \times p$ minors of $(p+1) \times m$ matrix over $\mathbb{Z}$ was obtained in this way, and it is worked out in [P-W]$_1$. The idea is to consider separately each "row" of the resolution (which are the pieces on which the differential is linear), and then to construct the resolution row by row.

A. Minors of a generic matrix.

One starts with the maps $\varphi : G \to F^*$, $\varphi^* : F \to G^*$ treating them as complexes non-zero in degrees -1 and 0. We have natural maps dual to each other

$$tr : R[-1] \to \varphi \otimes \varphi^* \ , \ ev : \varphi \otimes \varphi^* \to R[-1]$$

where R is a complex which is equal to 0 in degrees $\neq 0$ and to R in degree 0. Namely, fixing the basis $\{f_i\}$ in F and $\{g_j\}$ in G

$$tr(1) = \Sigma f_i^* \otimes f_i - \Sigma g_j \otimes g_j^*$$

Now one can construct the i'th row $X^i(\varphi)$ of the resolution of $p \times p$ minors of $m \times n$ matrix from the following double complex :

$$\begin{array}{ccccccc}
 & & \vdots & & \vdots & & \\
 & & \downarrow & & \downarrow & & \\
\cdots \to & \mathcal{K}_{s,t} & & \xrightarrow{\ \partial\ } & \mathcal{K}_{s-1,t} & \to \cdots & \\
 & \downarrow \delta & & & \downarrow \delta & & \\
\cdots \to & \mathcal{K}_{s,t+1} & & \xrightarrow{\ \partial\ } & \mathcal{K}_{s-1,t+1} & \to \cdots & \\
 & \downarrow & & & \downarrow & & \\
 & \vdots & & & \vdots & &
\end{array}$$

$$\mathcal{K}_{s,t} = \bigoplus_{\substack{|I|=s \\ |J|=t}} S_{I\setminus\boxed{m,i}\setminus J}\varphi \otimes S_{I\setminus\boxed{n,i}\setminus J}\varphi^* .$$

Here $S_{I\setminus\boxed{k,i}\setminus J}\varphi$ stands for the <u>skew</u> Schur functor associated with the shew - partition whose diagram can be expressed pictorially as follows :

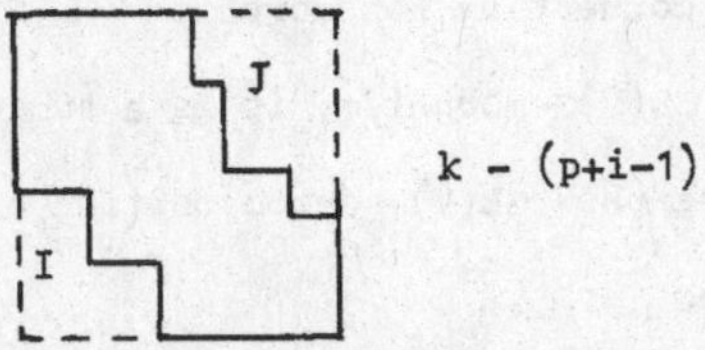

Skew Schur functor is defined in the following way.

Let $A \subset B$ be two partition $|A| = a$, $|B| = b$. Let D_A and D_B are situated in such a way that the first rows of D_A and D_B and the first columns of D_A and D_B are placed on the same primes. Define the skew-diagram $D_{B/A}$ as $D_B \setminus D_A$. Then $S_{B/A}(\varphi) = e(B/A)\,\varphi^{\otimes(b-a)}$. Here $e(B/A)$ as the following element in the group ring $\mathbb{Q}[\Sigma_{b-a}]$

$$e(B/A) = \sum_{\substack{r\in R(T_{B/A}) \\ c\in C(T_{B/A})}} (-1)^c r.c .$$

In this formula $T_{B/A}$ stands for $D_{B/A}$ filled with numbers $1,\dots,b-a$, and

$$R(T_{B/A}) = \{r\in\Sigma_{b-a}, r \text{ preserves the rows of } T_{B/A}\}$$
$$C(T_{B/A}) = \{c\in\Sigma_{b-a}, c \text{ preserves the columns of } T_{B/A}\} .$$

One has the following formula

$$S_{B/A}(\varphi) = \sum_{|C|=b-a} (C,A;B) S_C\varphi$$

where $(C,A;B)$ is the coefficient of S_C dans $S_A \otimes S_B$ and thus is described by Littlewood-Richardon-Schützenberger rule.

For simplifying the notation write $(I|m,i|J)$ instead of $S_{I\setminus\boxed{m,i}\setminus J}(\varphi)$ and $(I|n,i|J)^*$ instead of $S_{I\setminus\boxed{n,i}\setminus J}(\varphi^*)$.

In the case $m = n = p+1$ one easily gets the Gulliksen-Negard construction [G-N] . The next problem is to define the degree p map $X^i(\varphi) \to X^{i-1}(\varphi)$. This is done by identifying $X^i(\varphi)$ with the module $\mathrm{Ker}\,\delta / \mathrm{Ker}\,\delta \cap \mathrm{Im}\,\partial$ in the upper right corner of $\mathcal{K}_{s,t}$:

$$\begin{array}{ccc} (1|m,i|)\otimes(1|n,i|)^* & \xrightarrow{\partial} & (|m,i|)\otimes(|n,i|)^* \\ & & \downarrow \delta \\ & & (|m,i|\,1)\otimes(|n,i|\,1)^* . \end{array}$$

Now one defines the natural map :

$$(|m,i|)\otimes(|n,i|) \to (|m,i-1|)\otimes(|n,i-1|)$$

as the following composition : (here a_b (resp. (a_b^*)) stands for $S_{\underbrace{a\ldots a}_{b}}(\varphi)$ (resp. $S_{\underbrace{a\ldots a}_{b}}(\varphi^*)$)) .

$$\begin{array}{c} (|m,i|) \otimes (|n,i|) \\ \| \\ i_{(m-p-i+1)} \otimes i_{(n-p-i+1)} \\ \downarrow \text{Pieri} \\ K = (i-1)_{(m-p-i+1)} \otimes (i-1)^*_{(n-p-i+1)} \otimes 1_{(m-p-i+1)} \otimes 1_{(n-p-i+1)}^* \\ \downarrow 1 \otimes \mathrm{Tr} \otimes \mathrm{Tr} \\ K \otimes 1_{i-1} \otimes (1_{i-1})^* \otimes 1_{i-1} \otimes (1_{i-1})^* \\ \| \; \alpha \\ K \otimes 1_{i-1} \otimes (i-1^*) \otimes 1_{i-1} \otimes (i-1^*) \\ \downarrow \text{ multiplication} \\ (i-1)_{(m-p-i+1)} \otimes (i-1)_{(n-p-i+1)} \otimes 1_{m-p} \otimes (i-1^*) \otimes 1_{n-p}^* \otimes (i-1) \\ \downarrow \text{Pieri} \\ (i-1)_{(m-p-i)} \otimes (i-1)^*_{(n-p-i)} \otimes 1_{n-p} \otimes 1_{n-p}^* \\ \| \end{array}$$

$$\Vert$$

$$(|m,i-1|) \otimes (|n,i-1|)^* \otimes (|m,1|) \otimes (|m,1|)^*$$

$$\downarrow \beta$$

$$(|m,i-1|) \otimes (|n,i-1)^* .$$

Here the map α comes from the fact that $(\Lambda^{i-1}\varphi)^* = \delta_{i-1}\varphi^*$; the map β is just the composition : first the projection $\Lambda^{m-p}\varphi \otimes \Lambda^{n-p}\varphi^*$ onto the 0'th degree piece $\Lambda^p F \otimes \Lambda^p G$, and then embedding into R via $p \times p$ minors.

One proves in $[P\text{-}W]_1$ that this map induces a map of complexes $X^i(\varphi) \to X^{i-1}(\varphi)$ and that the total complex is a resolution of $p \times p$ minors. The main tool here is acyclicity lemma.

This construction seems promising, because it could be generalized to get the resolutions of the ideals I_λ described in [DC-E-P] , which doesn't seem possible to tackle using geometry.

<u>Example</u> : $\dim F = \dim G = p+2$, $\lambda = (p+1,p)$. The resolution of I_λ can be constructed as the total complex of the following double complex :

$$\begin{array}{ccccc}
R & \longrightarrow & X^{111}(\varphi) & \longrightarrow & X^{21}(\varphi) \\
 & & \uparrow & & \uparrow \\
 & & R & \longrightarrow & X^{1}(\varphi)
\end{array}$$

where for example $X^{21}(\varphi)$ is the unique i.e. $(0,0)$ - th homology of the following total complex

$$\begin{array}{ccccccc}
R \to \varphi\otimes\varphi^* + \varphi\otimes\varphi^* & \to & \varphi\otimes S_2\varphi\otimes\varphi^*\otimes S_2\varphi^* & \to & S_{21}\varphi\otimes S_{21}\varphi^* \\
\downarrow & & \downarrow & & \downarrow \\
R + R & \longrightarrow & \varphi\otimes\varphi^* + \varphi\otimes\varphi^* & \longrightarrow & S_2\varphi\otimes S_2\varphi^* + \Lambda^2\varphi\otimes\Lambda^2\varphi^* \\
 & & \downarrow & & \downarrow \\
 & & R & \longrightarrow & \varphi\otimes\varphi^* \\
 & & & & \downarrow \\
 & & & & R
\end{array}$$

For other details, in particular for a description of the differentials, see $[P\text{-}W]_1$ chap. 5.

B. A symmetric and anti-symmetric case.

Constructions analogous to the one outlined above are possible. In the anti-symmetric case we consider the map :

$$\mathrm{tr} : R[-1] \to S_2\varphi \text{ , } \mathrm{ev} : \Lambda^2\varphi \to R[-1]$$

where $\varphi : F \to F^*$ is our basic map (here $\varphi \cong \varphi^*$)

$$\mathrm{tr} : R[-1] \to \Lambda^2\varphi \text{ and } \mathrm{ev} : S_2(\varphi) \to R[-1] \text{ .}$$

One can now construct the rows similarly as in A , however the technical difficulties are greater there, so only small cases are worked out completely in [J-P-W] , namely $n-2p = 1,2,3$ for $2p$ - pfaffians and $n - p = 1,2$ for $p \times p$ symmetric minors.

III. CHARACTERISTIC FREE RESULTS.

The results of sections I,II are valid only in characteristic 0 because Bott's theorem is not true in characteristic $p > 0$. This makes the situation very difficult - in each case on has to choose carefully the appropriate $\mathbb{Z}$-form of a neaded functor. Before we proceed further, let us explain basic facts about $\mathbb{Z}$-forms.

Definition : Let A : (free R-Mod) → (free R-Mod) be a homogeneous functor defined for all R containing $\mathbb{Q}$ such that A commutes with base change. Then a $\mathbb{Z}$-form of A is by definition an homogeneous functor B : (free R-Mod) → (free R-Mod) which is defined for all R , commutes with base change, and such that B extends A .

Remark : Because of universality , A is determined by the homogeneous functor $A_{\mathbb{Q}}$: ($\mathbb{Q}$-Mod) → ($\mathbb{Q}$-Mod) , and B is determined by $B_{\mathbb{Z}}$: (free $\mathbb{Z}$-Mod) → (free $\mathbb{Z}$-Mod) . Thus finding a $\mathbb{Z}$-form of A boils down to finding a $\mathbb{Z}$-lattice inside of $A_{\mathbb{Q}}(F)$ (dim $F \geqslant$ deg A) which is invariant under the action of the envelopping algebra $U_{\mathbb{Z}}$ of the general linear group. Thus the $\mathbb{Z}$-forms correspond to the $\mathbb{Z}$-forms of the representation $A_{\mathbb{Q}}(F)$ in the sense of [V] . Thus one knows that for $A = S_I$ we have two extremal $\mathbb{Z}$-forms:

minimal and maximal ones, such that each other lies between them. We now describe these two $\mathbb{Z}$-forms : let $J = (j_1,\dots,j_q)$ be the conjugate partition with respect to $I = (i_1,\dots,i_r)$. The maximal $\mathbb{Z}$-form $L_I F$ is the image of the following map :

$$\begin{array}{ccc} \alpha_I : \Lambda^{j_1}F\otimes \dots \otimes \Lambda^{j_q}F & \longrightarrow & S_{i_1}F\otimes \dots \otimes S_{i_r}F \\ & {\scriptstyle\Delta}\searrow \quad \nearrow{\scriptstyle m} & \\ & F^{\otimes |I|} & \end{array}$$

where Δ stands for the diagonalisation, and m is the multiplication in each column of I . The minimal $\mathbb{Z}$-form $K_I F$ is the image of the map

$$\begin{array}{ccc} \beta_I : D_{i_1}F\otimes \dots \otimes D_{i_r}F & \rightarrow & \Lambda^{j_1}F\otimes \dots \otimes \Lambda^{j_q}F \\ & {\scriptstyle\Delta}\searrow \quad \nearrow{\scriptstyle m} & \\ & F^{\otimes |I|} & \end{array}$$

where Δ is again the diagonalisation, and m-the multiplication (compare [A],[W],$[\text{A-B-W}]_2$ for the properties of the $\mathbb{Z}$-forms L and K) . The inclusion $K_I F \to L_I F$ over $\mathbb{Z}$ is given by the following map :

$$K_I F \xhookrightarrow{\text{inclusion}} \Lambda^{j_1}F\otimes \dots \otimes \Lambda^{j_q}F \longrightarrow L_I F .$$

Note, that we have another inclusion (over $\mathbb{Z}$) $L_I F \hookrightarrow K_I F$:

$$L_I F \to S_{i_1}F\otimes \dots \otimes S_{i_r}F \to D_{i_1}F\otimes \dots \otimes D_{i_r}F \xrightarrow{\beta_I} K_I F .$$

The composition of these maps is easily seen to be the multiplication by the scalar $\prod_{(i,j)} h_{ij}$, where h_{ij} is the hook length of element (i,j) in D_I (compare [Bo]) .

A. Minors of generic matrix.

$p \times p$ minors of $p \times m$ matrix ([E-N]) and $p \times p$ minors of $(p+1) \times m$ matrix ($[\text{A-B-W}]_1$) have characteristic - free resolutions. In the Eagon-Northcott case one uses divided powers. The submaximal case is more complicated - one first construct two rows separately. The first row $X^1(\varphi)$ is the homology of the complex

$$0 \longrightarrow \Lambda^{m-p-1}\varphi \xrightarrow{\partial} \Lambda^{m-p}\varphi\otimes\varphi^* \xrightarrow{\delta} \Lambda^{m-p-1}\varphi \longrightarrow 0$$

which turns out to be characteristic free. This is rather subtle, because the map δ

involves coefficients (the multiplication in divided powers algebra) . The second row $X^2(\varphi)$ is $S_{m-p-1,m-p-1}\varphi$. The interesting thing is that the connecting map $X^2(\varphi) \to X^1(\varphi)$ is not natural.

In $[A-B-W]_1$ the natural, characteristic free candidate for the first row is constructed in the general case. The higher rows are still a mystery .

We should mention also a paper of Roberts $[R]_2$ in which he attempted to construct the resolution of minors over $\mathbb{Z}$ without decomposing it into rows.

B. The symmetric and antisymmetric cases.

In the symmetric case, the resolution of length 3 is characteristic free [J] . Here one needs a subtle theorem on the quadratic forms in characteristic 2 to get the result.

In the antisymmetric case the resolution of length 3 is well known to be characteristic free $[B-E]_2$. Also the length 6 case is independent of characteristic $[P]_2$. There is some trouble with characteristic 2 there . The reason is that there exists a $\mathbb{Z}$-relation among n-2-order Pfaffians which is <u>not</u> a consequence of the Laplace type relations. Therefore the use of some special $\mathbb{Z}$-form for the first syzygies is necessary.

IV. BIBLIOGRAPHY.

For the reader's convenience, we have added here references which did not appear in the text.

[A] K. Akin, Thesis, Brandeis (1980).

$[A-B-W]_1$ K. Akin, D.A. Buchsbaum, J. Weyman, Resolutions of determinantal ideals : the submaximal minors, Adv. in Math. 39 (1981), 1-30.

$[A-B-W]_2$ Schur functors and Schur complexes, Adv. in Math. 44 (1982),207-278.

[Bo] H. Boerner, Representations of Groups, North Holland Pub. Co. 1970.

Burch L., On ideals of finite homological dimension in local rings, Proc. Camb. Phil. Soc. 64 (1968), 949-952.

$[B-E]_1$ D.A. Buchsbaum, D. Eisenbud, What makes a complex exact ? J. of Alg. 25 (1973), 259-268.

$[B-E]_2$ Algebra structure for finite free resolutions and some structure theorems for ideals of codimension three, Amer. J. Math. 99 (1977), 447-485.

$[B-E]_3$ Generic free resolutions and a family of generically perfect ideals, Adv. in Math. 18 (1975), 245-301.

E. Cartan, S. Eilenberg, Homological algebra, Princeton (1952).

S. Eilenberg, Homological dimension and syzygies, Annals of Math., 64 (1956), 328-336.

[DC-E-P] C. De Concini, D. Eisenbud, C. Procesi, Young diagrams and determinantal ideals, Inv. Math., 56 (1980), 129-165.

[E-N] J.A. Eagon, D.G. Northcott, Ideals defined by matrices and a certain complex associated with them, Proc. Roy. Soc. London Ser. A 269 (1962), 188-204.

[G-N] T.H. Gullikeen, O.G. Negard, Un complexe résolvant pour certains idéaux déterminantiels, C.R. Acad. Sci. Paris Sér. A 274 (1972), 16-19.

[H] D. Hilbert, Uber die Theorie der algebraischen Formen, Math. Ann., 36 (1890).

[J] T. Józefiak, Ideals generated by minors of a symmetric matrix, Comment. Math. Helv., 53 (1978), 595-607.

[J-P] T. Jozefiak, P. Pragacz, Syzygies de Pfaffians, C.R. Acad. Sci. Paris, 287 (1978), 89-91.

[J-P-W] T. Józefiak, P. Pragacz, J. Weyman, Resolutions of determinantal varieties and tensor complexes associated with symmetric and anti-symmetric matrix, Astérisque, 87-88 (1981), 109-189.

[K] G. Kempf, On the collapsing of homogenous bundles, Inv. Math., 37 (1976), 229-239.

$[L]_1$ A. Lascoux, Syzygies des varietiés déterminantales, Adv. in Math. 30 (1978), 202-237.

$[L]_2$ Syzygies pour les mineurs de matrices symétriques, Preprint, Paris (1977).

[L-S] A. Lascoux, M. P. Schützenberger, Le Monoîde Plaxique, Napoli (1978), Quaderni della Ricerca Scientifica n. 109. Non-commutative structures in Algebra and Geometric Combinatorics, Roma 1981.

F.S. Macaulay, The Algebraic Theory of Modular Systems, Cambridge Tracts vol. 19 (1916).

[Mc] I.G. Macdonald, Symmetric Functions and Hall Polynomials, Oxford Math. Monogr., (1979).

[N] H.A. Nielsen, Tensor functors of complexes, Aarhus University, Preprint N° 15 (1977/1978).

[P-S] C. Peskine, L. Szpiro, Dimension projective finie et cohomologie locale, I.H.E.S. Publ. Math., 42 (1973), 323-395.

[Po] K.Y. Poon, Thesis, Minnesota (1973).

$[P]_1$ P. Pragacz, Thesis, Toruń (1981).

$[P]_2$ Characteristic free resolution of n-2-order Pfaffians of $n \times n$ anti-symmetric matrix, J. of Alg. 78 (1982), 386-396.

$[P\text{-}W]_1$ P. Pragacz, J. Weyman, Complexes associated with trace and evaluation ; another approach to the Lascoux's resolution, Preprint, Toruń (1982).

[P-W]$_2$ On the generic free resolutions, in preparation.

[R]$_1$ P. Roberts, A minimal free complex associated to the minors of matrix, Preprint.

[R]$_2$ On the construction of the generic resolutions of determinantal ideals, Asterisque 87-88, (1981), 353-378.

[S] Study, Methoden der Theorie der Ternaren Formen, Leipzig 1880.

[Sch] A. Schönflies : Bemerkung zu Hilbert's Theorie der algebraischen Formen, Göttinger Nachrichten (1891), 339-399.

[Se] Segre B. Enzyklopedie der Mathematischen Wissenschaften III,Geometrie, teil 2.

[W] J. Weyman, Thesis, Brandeis (1980).

[V] D.N., Verma, The role of affine groups in the representation theory of algebraic Chevalley groups and their Lie algebras, in "Lie Groups and their Representations", pp. 653-705, Akadémiai Kiadó Budapest (1975).

Département de Mathématique
U.L.P. Strasbourg

Institute of Mathematics
Polish Academy of Sciences

L'ELIMINATION LINEAIRE DANS LES CORPS GAUCHES

Paul Van Praag
Université de Mons-Hainaut
Institut de Mathématiques
Avenue Maistriau, 15
7000 MONS, BELGIQUE

1. Introduction

L'existence d'une solution x du système d'équations

$$\begin{cases} ax = b \\ cx = d \end{cases} \qquad (^0)$$

est équivalente pour tout corps commutatif à l'expression sans quantificateurs :

"((a≠0 ou c≠0) et ad-bc = 0) ou (a=0 et b=0) et c=0 et d=0)"

Bien qu'il soit question de corps, cette dernière expression est écrite dans "le langage naturel des anneaux" (voir 2.) : on y utilise des signes nécessaires pour écrire en théorie des anneaux, on n'y utilise pas l'expression x^{-1}.

Un cas particulier du problème traité ici est de rechercher les corps non commutatifs pour lesquels existe, lorsque $ac \neq ca$, une telle expression équivalente à l'existence d'une solution pour $(^0)$. Ce système possède bien sûr une solution si et seulement si

$$c\,a^{-1}\,b = d\,,$$

mais cette formule n'est pas écrite dans le langage naturel des anneaux.

La recherche pour un langage et une théorie donnée, des formules qui sont équivalentes à des formules sans quantificateurs est une branche de la logique mathématique qui généralise la théorie de l'élimination en algèbre. On en trouvera un historique et des motivations par exemple dans [5] et dans les ouvrages [2] et [13].

Nous aurons besoin des éléments suivants :

2. Le langage naturel des anneaux et l'élimination des quantificateurs

Le langage usuel L_a de la théorie des anneaux est formé des signes suivants :

$+ , . , - , =$

$0 , 1$ (les constantes)

$x_1 , x_2, \ldots$ (les variables)

$\wedge , \neg , \exists$ (les signes logiques)

$(, , ,)$ (les signes impropres) .

Les termes sont définis par induction :

1°) les constantes et les variables sont des termes

2°) si t_1 et t_2 sont des termes, alors t_1+t_2 , $t_1.t_2$ et $-t_1$ sont des termes.

Les formules sont définies par induction :

1°) si t_1 et t_2 sont des termes, alors $t_1=t_2$ est une formule (formule atomique)

2°) si φ est une formule, alors $\neg\varphi$ est une formule

3°) si φ et ψ sont des formules, alors $\varphi \wedge \psi$ est une formule

4°) si φ est une formule et x une variable, alors $(\exists x)\varphi$ est une formule.

On fait les abréviations suivantes :

On abrège

$\neg (t_1=t_2)$ en $t_1 \neq t_2$

$\neg (\neg\varphi \wedge \neg \psi)$ en $\varphi \vee \psi$

$\neg ((\exists x)(\neg \varphi))$ en $(\forall x) \varphi$

$\neg\varphi \vee \psi$ en $\varphi \rightarrow \psi$

$(\varphi \rightarrow \psi) \wedge (\psi \rightarrow \varphi)$ en $\varphi \leftrightarrow \psi$

On définit les notions de variable libre et de variable liée. Par exemple, dans la formule

$(\exists x) (x+y = 0)$,

x est lié et y est libre. On dit que φ est une formule en les variables libres $x_1,\ldots,x_n$ et on écrit $\varphi(x_1,\ldots,x_n)$ si les variables libres de φ appartiennent à $\{x_1,\ldots,x_n\}$.

Un énoncé (ou une sentence) est une formule qui ne contient aucune variable libre.

La théorie T des anneaux est l'ensemble des énoncés qui sont les axiomes usuels :

$(\forall x)(\forall y)(\forall z)(x+(y+z)) = ((x+y)+z)$ etc

et de tous les énoncés qui s'en déduisent à l'aide de règles bien précises telles que par exemple :

Si φ et $(\varphi \to \psi)$ sont des éléments de la théorie, alors ψ est aussi un élément de la théorie

Un théorème est un élément de la théorie.
En ajoutant des axiomes on obtient des théories particulières des anneaux. Les formules $\varphi(x_1,\ldots,x_n)$ et $\psi(x_1,\ldots,x_n)$ en les variables libres $x_1,\ldots,x_n$ sont <u>équivalentes</u> (dans T) si

$(\forall x_1)\ldots(\forall x_n)\ (\varphi(x_1,\ldots,x_n) \leftrightarrow \psi(x_1,\ldots,x_n))$

est un théorème de T.

On montre que dans T toute formule atomique est équivalente à une formule de la forme

$p(x_1,\ldots,x_n) = 0$ (1)

où p est un polynôme à cœfficients dans Z en les variables non commutatives $x_1,\ldots,x_n$.

On montre que toute formule sans quantificateurs de L_a est équivalente à une conjonction (resp. disjonction) de disjonctions (resp. conjonctions) de formules (1) ou de négation de telles formules (une formule sans quantificateurs de L_a n'est en général pas équivalente dans T à une formule atomique ou à sa négation. Par exemple la formule suivante : $((a=0) \to (b=0))$).

Une structure pour le langage L_a est la donnée de

$\mathbb{A} = (A, +_A, -_A, \cdot_A, 0_A, 1_A)$ (2)

où A est un ensemble non vide, $+_A$ et $\cdot_A$ sont des applications de $A \times A$, dans A , $-_A$ est une transformation de A et 0_A, 1_A sont des élements de A .

Soient $\varphi(x_1,\ldots,x_n)$ une formule de L_a en les variables libres $x_1,\ldots,x_n$; $\mathbb{A}$ une structure pour L_a, et $a_1,\ldots,a_n \in A$.
On dit que $\mathbb{A}$ satisfait φ en $A_1,\ldots,a_n$ et on écrit

$\mathbb{A} \models \varphi(a_1,\ldots,a_n)$,

si $\varphi(a_1,\ldots,a_n)$ est "vraie" lorsque l'on remplace $x_1,\ldots,x_n$ par $a_1,\ldots,a_n$; + par $+_A$; etc .
On dit que $\mathbb{A}$ est un modèle pour T si et seulement si

$A \models \varphi$

pour tout élément φ de T .

Soit T_1 une théorie d'anneaux.

On dit que T_1 a l'élimination des quantificateurs (EQ) si pour toute formule $\varphi(x_1,\ldots,x_n)$ de L_a en les variables libres $x_1,\ldots,x_n$, il existe une formule sans quantificateurs $\theta(x_1,\ldots,x_n)$ équivalente à $\varphi(x_1,\ldots,x_n)$ dans T_1 .

Soit $\mathcal{A}$ une classe d'anneaux. On dit que $\mathcal{A}$ a l'EQ si pour toute formule $\varphi(x_1,\ldots,x_n)$ de L_a , il existe une formule sans quantificateurs $\theta(x_1,\ldots,x_n)$ telle que

$$\mathbb{A} \models (\forall x_1) \ldots (\forall x_n) \, (\varphi(x_1,\ldots,x_n) \leftrightarrow \theta(x_1,\ldots,x_n))$$

pour tout $\mathbb{A} \in \mathcal{A}$.

Le premier exemple fut le <u>théorème de TARSKI</u> : La classe (ou la théorie) des corps algébriquement clos à l'EQ.

Voici une conséquence : soit φ un énoncé de L_a .

Il est équivalent sur tout corps algébriquement clos à une disjonction de conjonctions de formules atomiques ou de négations de telles formules Mais les formules atomiques qui sont des énoncés sont de la forme

$$z = 0$$

où $z \in Z$.

Le théorème veut donc dire qu'il existe un nombre fini de $z_i \in Z$ tels que φ soit équivalent dans tout corps algébriquement clos à une disjonction de formules du type

$$z_1 = 0 \wedge \ldots \wedge z_s = 0 \wedge z_{s+1} \neq 0 \wedge \ldots \wedge z_t \neq 0 \, .$$

Or on peut vérifier en un nombre fini d'étapes si une telle formule est vraie ou fausse. On dit alors que la théorie des corps algébriquement clos est <u>décidable</u>.

Un autre exemple de théorie prouvée décidable par TARSKI à l'aide de l'EQ est la théorie des corps réellement clos.

Ici, on change de langage : on ajoute le signe $<$. On montre que, pour le nouveau langage, la théorie des corps réellement clos à l'EQ et est donc décidable. On en déduit que la géométrie plane élémentaire (convenablement définie) l'est aussi (voir exemple [13]).

Il est instructif pour la suite de l'exposé d'indiquer les grandes lignes de la preuve du théorème de TARSKI pour les corps algébriquement clos : on montre tout d'abord qu'il suffit d'éliminer $\forall$ dans les formules de la forme

$$(\forall x)\ \theta(x,x_1,\ldots,x_n)$$

où θ est sans quantificateurs.
On se ramène aux formules du type

$$(\forall x)\ (p_1(x,x_1,\ldots,x_n) = 0 \vee \ldots \vee p_s(x,x_1,\ldots,x_n) = 0$$
$$\vee\ q_1(x,x_1,\ldots,x_n) \neq 0 \vee \ldots \vee q_t(x,x_1,\ldots,x_n) \neq 0) \qquad (^3)$$

où les p_i et les q_i sont des polynômes.
Soit K un corps commutatif et fixons $x_1,\ldots,x_n \in K$.
On note $p(x,x_1,\ldots,x_n)$ le produit des $p_i(x,x_1,\ldots,x_n)$ et q le plus grand commun diviseur des $q_i(x,x_1,\ldots,x_n)$.
Dès lors

$$p_1(x,x_1,\ldots,x_n) = 0 \vee \ldots \vee p_s(x,x_1,\ldots,x_n) = 0$$

est une formule équivalente à

$$p(x,x_1,\ldots,x_n) = 0\ ,$$

et

$$q_1(x,x_1,\ldots,x_n) = 0 \wedge \ldots \wedge q_t(x,x_1,\ldots,x_n) = 0$$

est équivalente à

$$q(x,x_1,\ldots,x_n) = 0\ .$$

Donc $(^3)$ est équivalente à

$$(\forall x)\ (q(x,x_1,\ldots,x_n) \neq 0 \text{ ou } p(x,x_1,\ldots,x_n) = 0)\ ,$$

c'est-à-dire

$$(\forall x)\ (q(x,x_1,\ldots,x_n) = 0 \rightarrow p(x,x_1,\ldots,x_n) = 0) \qquad (^4)$$

Soit maintenant K algébriquement clos. Dès lors $(^4)$ est équivalente à

$$q \mid p^m$$

où m est un entier quelconque mais $\geq \deg q$.
En fait, puisque $q \mid q_i$, on peut prendre $m = \max \deg q_i$.
Mais $q \mid p^m$, c'est-à-dire que p^m appartient à l'idéal engendré par q, c'est-à-dire $idl(q_1,\ldots,q_t)$. Le problème revient donc à rechercher les $h_i \in K[x]$ pour lesquels

$$p^m = h_1q_1 + \ldots + h_tq_t\ . \qquad (^5)$$

A l'aide de l'algorithme d'Euclide, on montre que l'on peut borner les degrés des h_i en fonction de ceux des q_i et de p.
Les cœfficients des h_i sont alors solutions du système d'équations linéaires obtenu en égalent les cœfficients de x dans les deux membres de (^{5}). L'existence du déterminant et le théorème de KRONECKER-ROUCHE-CAPELLI fournissent alors une formule sans quantificateurs équivalente à l'existence d'une solution pour ce système. □

On a recherché des classes d'anneaux qui admettent l'EQ. Citons comme résultats marquants :

Les seuls corps commutatifs qui admettent l'EQ (pour L_a) sont les corps algébriquement clos et les corps finis [12].

Les seuls anneaux de caractéristique zéro qui admettent l'EQ sont les corps commutatifs algébriquement clos [3].

Une classification des anneaux semi-simples qui admettent l'EQ [4].

VAN DEN DRIES ([16]) a introduit la définition suivante :

Une classe d'anneaux $\mathcal{A}$ a l'<u>élimination linéaire</u> (EL) si pour toute formule existentielle Ψ_{mn} (en les variables libres a_{ij} et b_i) qui exprime le système

$$\sum_{j=1}^{n} a_{ij}x_j = b_i \quad (i=1,\dots,m) \tag{6}$$

possède une solution, il existe une formule sans quantificateurs de L_a qui est équivalente à Ψ_{mn} dans chaque anneau de $\mathcal{A}$.
La notion plus faible d'EL-n, introduite par Françoise POINT ([14], [15]), s'est révélée utile. En particulier, on dit que la classe $\mathcal{A}$ a l'élimination linéaire -2 (EL-2) s'il existe une formule sans quantificateurs de L_a équivalente à Ψ_{21} dans tous les anneaux de $\mathcal{A}$.

3. Les corps gauches qui admettent l'élimination linéaire

3.1. Le premier résultat fut prouvé par Françoise POINT dans sa thèse [15] :

<u>Théorème</u> : Tout corps qui a l'élimination linéaire -2 est de rang fini sur son centre.
Nous donnons ici la preuve originale et une modification de cette preuve due à Maurice BOFFA ([6]) et qui nécessite beaucoup moins de logique.

[6] et [7] posaient en outre les trois questions suivantes :

Existe-t-il des corps non commutatifs qui ont l'EL-2 ?

Existe-t-il des corps non commutatifs qui ont l'EL ?

Qu'en est-il du corps des quaternions ?

Tout d'abord, pour la preuve originale, quelques notions et remarques :

1) Soient $\mathbb{A} \subset \mathbb{B}$ des structures pour L_a.

On dit que $\mathbb{A}$ est une <u>sous-structure élémentaire</u> de $\mathbb{B}$, ou que $\mathbb{B}$ est une <u>extension élémentaire</u> de $\mathbb{A}$, et on écrit

$$A \prec \mathbb{B}$$

si pour toute formule $\varphi(x_1,\dots,x_n)$ de L_a en les variables libres $x_1,\dots,x_n$ et pour tout $a_1,\dots,a_n \in A$, on a

$$\mathbb{A} \models \varphi(a_1,\dots,a_n) \quad \text{ssi} \quad \mathbb{B} \models \varphi(a_1,\dots,a_n).$$

Soient $\mathbb{A}$ et $\mathbb{B}$ des anneaux et $\mathbb{A} \prec \mathbb{B}$.
Si $\mathbb{A}$ a l'EL-2, alors $\mathbb{B}$ aussi (en effet, il existe une formule sans quantificateurs $\theta(a,b,c,d)$ de L_a, telle que

$$\mathbb{A} \models (\forall a)(\forall b)(\forall c)(\forall d)((\exists x)(ax=b \land cx=d) \leftrightarrow \theta(a,b,c,d))\ .$$

Voici un procédé pour "construire" des extensions élémentaires : les ultrapuissances. Soit $A_{i\in N}$ une famille de copies isomorphes de A, et $\mathcal{U}$ un ultrafiltre sur N (c'est-à-dire, est un ensemble non vide de parties de N tel que

$$\emptyset \notin \mathcal{U}$$

$$x,y \in \mathcal{U} \to x \cap y \in \mathcal{U}$$

$$x \in \mathcal{U} \to N - x \in \mathcal{U}) .$$

On prouve qu'il existe des ultrafiltres (les ultrafiltres non principaux) qui contiennent l'ensemble des parties cofinies de N.
Sur le produit $\prod_{i\in N} A_i$, on définit la relation d'équivalence

$$(a_i) \sim (b_i)$$

ssi $\{\, i \mid a_i = b_i \,\} \in \mathcal{U}$.

On prouve que $\prod_i A_i / \sim$ est un anneau, et que de plus :

$$A \prec \prod_i A_i / \sim .$$

En particulier, une ultrapuissance de corps est un corps puisque un corps est un anneau qui vérifie l'énoncé

$(\forall x)\ ((x \neq 0) \rightarrow (\exists y)\ (xy = 1 \wedge yx = 1))$.

2) Si D est un corps de rang infini sur son centre, il possède une extension élémentaire qui comprend des éléments libres sur le corps premier de D. Montrons cela à l'aide des ultrapuissances.

Posons $P = Z$ si la caractéristique de D est nulle et $P = F_p$ si elle est p. Enumérons les éléments de $P\langle x,y\rangle$, l'algèbre des polynômes en les variables non commutatives x et y à cœfficients dans P :

$$P\langle x,y\rangle = \{0, p_1, p_2, \ldots \} .$$

Il existe deux suite (x_i) et (y_i) telles que

$$\begin{array}{lll} p_1(x_1,y_1) \neq 0 & p_1(x_2,y_2) \neq 0 \ \ldots & p_1(x_k,y_k) \neq 0 \ \ldots \\ & p_2(x_2,y_2) \neq 0 & p_2(x_k,y_k) \neq 0 \\ & & \vdots \\ & & p_k(x_k,y_k) \neq 0 \end{array}$$

(Pour tout k , il existe de tels x_k, y_k, sinon $p_1 \ldots p_k$ serait une identité sur D , ce qui contredit l'hypothèse sur D).

Soient $\mathcal{U}$ un ultrafiltre non principal sur N ,

$$\overline{D} = \prod_{i \in N} D_i / \sim ,$$

et x et y les classes dans $\overline{D}$ de $(x_1,x_2,\ldots)$ et de $(y_1,y_2,\ldots)$. Dès lors x et y sont libres sur P , sinon ils annulleraient un élément non nul p_ℓ de $P\langle x,y\rangle$. Mais $p_\ell(x,y)=0$ s'écrit aussi

$$(p_\ell(x_1,y_1), p\ (x_2,y_2), \ldots) \sim (0,0,\ldots) ,$$

c'est-à-dire

$$\{i \mid p_\ell(x_i,x_i) = 0\} \in \mathcal{U} ,$$

Mais les x_i et y_i sont justement définis de façon à ce qu'un tel ensemble de gauche soit fini. (En agrandissant le langage on peut montrer que u et v peuvent être pris libres sur le centre de D).

<u>Preuve (originale) du théorème</u> : Supposons que D soit de rang infini sur son centre et admette l'EL-2, et soit $\theta(a,b,c)$ une formule sans quantificateurs de L_a équivalente sur D à

$$(\exists x)\ (ax=b) \wedge bx=c)$$

$\theta(a,b,c)$ est équivalente à une conjonction de disjonctions de formules du type

$$p_i(a,b,c) = 0$$

ou $p_i(a,b,c) \neq 0$,

où $0 \neq p_i \in P\ x,y,z$.

Soit (par (2)) une extension élémentaire D_1 de D qui comprenne des éléments u et v , libres sur P .

On a $\theta(u,uv,uv^2)$ puisque $u.v=uv$ et $uv.v=uv^2$. Puisque u et v sont libres sur P , u, uv et uv^2 le sont aussi, donc pour tout i :

$$p_i(u,uv,uv^2) \neq 0 .$$

Mais u, uv, uv^3 est aussi une famille libre sur P , donc $p_i(u,uv,uv^3) \neq 0$. Donc les triples (u,uv,uv^2) et (u,uv,uv^3) désanullent les mêmes p_i . Donc $\theta(u,uv,uv^3)$, et dès lors

$$(\exists z)\ (uz = uv \ \wedge\ uvz = uv^3) ,$$

ce qui contredit la liberté de u et de v . □

<u>Preuve de BOFFA</u> : Soient D, θ et les p_i comme dans la preuve originale. Puisque X et Y sont libres sur P , les triples (X,XY,XY^2) et (X,XY,XY^3) le sont aussi.

Donc le polynôme

$$F = XY(Y-1) \prod_i p_i(X,XY,XY^2)\ p_i(X,XY,XY^3)$$

est non nul. On va prouver que F est une identité sur D (et donc D est de rang fini sur son centre). Si ce n'était pas le cas, il existerait x et $y \in D$ tels que

$$F(x,y) \neq 0 \qquad (^7)$$

Mais $\theta(x,xy,xy^2)$ puisque $x.y = xy$ et $xy.y = xy^2$. L'inégalité $(^7)$ implique que tous les $p_i(x,xy,xy^2)$ et les $p_i(x,xy,xy^3)$ sont non nuls.

Dès lors (x,xy,xy^2) et (x,xy,xy^3) désanullent les mêmes p_i , donc

$$\theta(x,xy,xy^3) ,$$

c'est-à-dire

$$(\exists z)\ (xz=xy \ \text{ et } \ xyz=xy^3) ,$$

ce qui contredit $(^7)$. □

3.2. L'auteur de ces lignes prouva comme ceci l'EL-2 pour les quaternions :

supposons que dans le systèmes $(^0)$, a,b,c et d prennent leurs valeurs dans un corps de rang 4 sur son centre et que $ac \neq ca$. Ce système est alors équivalent à ceux-ci :

$$\begin{cases} cax = cb \\ acx = ad \end{cases}$$

$$\begin{cases} ca-ac)^2x = (ca-ac)(cb-ad) \\ cx = d \end{cases}$$

lequel possède une solution si et seulement si

$$c(ca-ac)^{-2}(ca-ac)(cb-ad) = d$$

Mais ac-ca est un quaternion imaginaire pur, donc $(ca-ac)^2$ est central (identité de WAGNER), donc l'existence d'une solution de $(^0)$ est équivalente à

$$c(ca-ac)(cb-ad) - d(ca-ac)^2 = 0 \qquad (^8)$$

ce qui est une formule sans quantificateurs de L_a . □

A partir de cette preuve, le même auteur étendit ([8], prop. 2; [17], appendice) ce résultat à la classe des corps de rang $\leq n$ sur leur centre quelque soit n (et par cette preuve, l'existence d'une solution de $(^0)$ est équivalente sur tous les corps de rang 4 à

$$(a^2c-ca^2)(ad-cb) - (ac-ca)(a^2d-cab) = 0 \qquad (^9)).$$

3.3. Suite à ces résultats BOFFA et MICHAUX prouvaient ([8], prop. 3) que pour chaque entier n, la classe des corps de dimension $\leq n$ sur leur centre à l'élimination des inverses (voir ci-dessous) et donc l'EL.

Définition : soit $L_c = L_a \cup \{\ ^{-1}\}$ le langage naturel des corps et posons $0^{-1}=0$ (cette convention à première vue surprenante pour un non logicien ne contredit aucun axiome des corps). On dit qu'une classe de corps a <u>l'élimination des inverses</u> (EI) si pour chaque formule sans quantificateurs de L_c , il existe une formule sans quantificateurs de L_a qui lui est équivalente dans chaque corps de cette classe. Une classe de corps qui a l'EI a bien sûr l'EL (GAUSS) et dès lors

<u>Théorème</u> [8] : <u>Pour tout corps D, les conditions suivantes sont équivalentes</u>

1) <u>D est de rang fini sur son centre</u>

2) <u>D a l'élimination linéaire -2</u>

3) D a l'élimination linéaire

4) D a l'élimination des inverses.

(Pour les algèbres non associatives, cela semble plus difficile : André LEROY a prouvé [11] que l'algèbre des octaves de CAYLEY a l'EL-2, mais la question de savoir si par exemple elle a aussi l'EL est à la connaissance de l'auteur encore ouverte).

Après la rédaction de [8], BOFFA remarqua (lettre à l'auteur) que l'implication 1) → 4) pouvait se déduire d'un résultat de la théorie des modèles (montrant que si C est une classe élémentaire de corps (c'est-à-dire formée de tous les corps qui sont modèles d'une même théorie) telle que tout isomorphisme de sous-anneaux d'éléments de C peut s'étendre en un isomorphisme de sous-corps de ces éléments, alors C a l'EI) et du fait que tout sous-anneau d'un corps de dimension finie sur son centre est un anneau de ORE.

Mais l'intérêt de la preuve de BOFFA-MICHAUX dans [8] est qu'elle est effective et qu'elle dit un peu plus : elle contient et repose sur le lemme suivant dont on déduit (effectivement) que tout sous-anneau d'une algèbre algébrique sur son centre et sans diviseurs de zéro est un anneau de ORE :

Lemme 1 : Pour tout n, il existe des polynômes $p_1,\dots,p_n$, $q_1,\dots,q_n$ en les variables non commutatives X et Y et à coefficients entiers, tels que si A est une algèbre sur K sans diviseur de zéro, x et y sont des éléments non nuls de A, y est algébrique de degré $\leq n$ sur K, alors il existe $i \in \{1,\dots,n\}$ pour lequel

$$p_i(x,y) \neq 0$$

et $p_i(x,y)x = q_i(x,y)y$.

Preuve : Notons $S_m(X_1,\dots,X_m)$ le polynôme standard

$$\Sigma \pm X_{1\sigma} \dots X_{m\sigma}$$

où σ parcourt le groupe des permutations de $\{1,\dots,m\}$ et le signe $\pm$ dépend de la parité de σ.

On pose

$$S_{m,\Gamma}(x,y) = S_m(xy^{\Gamma}, \dots, xy^{\Gamma+m-1}).$$

Groupons les termes de $S_m(X_1,\dots,X_m)$ sui se terminent par X_1 :

$$S_m(X_1,\dots,X_m) = (\sum_{\theta} \pm X_{2\theta} \dots X_{m\theta})X_1$$

$$+ \sum_{\ell=2}^{m} (\sum_{\eta} \pm X_{1\eta} \dots \hat{X}_{\ell\eta} \dots X_{m\theta})X_{\ell}$$

où θ et η parcourent respectivement les groupes de permutations de $\{2,\ldots,m\}$ et de $\{1,\ldots,\hat{\ell},\ldots,m\}$ (le chapeau sur un terme signifiant son ommission). En remplaçant X_j par $x\,y^{\Gamma+j-1}$, on en déduit que

$$S_{m,\Gamma}(x,y) = (\pm S_{m-1,\Gamma+1}(x,y)x + Q_{m,\Gamma}(x,y)y)\,y^{\Gamma}$$

$$\text{où } Q_{m,\Gamma}(x,y) = \sum_{q=1}^{m-1} (\pm S_{m-1}(xy^{\Gamma},\ldots,\widehat{xy^{\Gamma+q}},\ldots,xy^{\Gamma+m-1})\,xy^{q-1}$$

Si y est algébrique de degré $\leq n$ sur K, alors les $n+1$ éléments $x,xy,\ldots,xy^n$ sont linéairement indépendants sur K, donc

$$\begin{aligned} 0 &= S_{n+1}(x,xy,\ldots,xy^n) \\ &= S_{n+1,o}(x,y) \\ &= \pm S_{n,1}(x,y)\,x + Q_{n+1,o}(x,y)\,y\,. \end{aligned}$$

Si $S_{n,1}(x,y) \neq 0$, alors on pose

$$p_1(x,y) = \pm S_{n,1}(x,y)$$

et $q_1(x,y) = Q_{n+1,o}(x,y)$.

Si $S_{n,1}(x,y) = 0$, alors

$$\pm S_{n-1,2}(x,y) + q_2(x,y)\,y = 0\text{ , etc.}$$

On finit par s'arrêter puisque

$$S_{1,n}(x,y) = xy^n \neq 0\,. \quad \square$$

4. On déduit du lemme que si D est un corps de rang fini sur son centre, il existe un procédé effectif pour mettre toute expression rationnelle portant sur des éléments de D sous la forme $p^{-1}q$ où p et q sont des expressions polynômiales.

Une suggestion de BOFFA d'appliquer ce procédé au calcul de représentants de déterminants de DIEUDONNE amena l'auteur de ces lignes à remarquer que ce déterminant pouvait servir de cadre aux résultats précédents sur l'EL et relier par exemple les premiers membres des égalités ([8]) et ([9]) [18].

Rappelons que si D est un corps et $A \in M_n(D)$, le déterminant de DIEUDONNE de A, det A, prend ses valeurs dans $\{\bar{o}\} \cup D^{*n}/[D^*,D^*]$ où D^* est le groupe multiplicatif de D, $[D^*,D^*]$ est le sous-groupe de D^* engendré par les commutateurs $aba^{-1}b^{-1}$ $(a,b \in D^*)$ et $\bar{o}$ est un zéro adjoint au groupe $D^*/[D^*,D^*]$ (voir [9],[1] ou [10]).

Si D est commutatif, on retrouve le déterminant usuel.

La matrice A est inversible si $\det A \neq \overline{o}$.
Si $a \in D$, on pose $\overline{a} = a\,[D^*,D^*]$. On prouve par exemple que si $a \neq o$, alors

$$\det \begin{pmatrix} a & b \\ c & d \end{pmatrix} = \overline{ad - aca^{-1}b}$$

et que

$$\det \begin{pmatrix} 0 & b \\ c & d \end{pmatrix} = \overline{-cb}$$

Plus généralement, un représentant de det A dans $D^*/[D^*,D^*]$ est une expression rationnelle en les A_{ij} et dès lors, si D est de rang fini sur son centre, est égal à une expression de la forme $p^{-1}q$ où p et q sont des expressions polynômiales en les A_{ij}. Ce qui permet d'exprimer l'EL pour D à l'aide de numérateurs et de dénominateurs de représentants du déterminant de DIEUDONNE (puisqu'une matrice à cœfficients dans un corps gauche a aussi un rang et, comme dans le cas commutatif, le système (^{4}) y possède un zéro si et seulement si le rang de la matrice (a_{ij}, b_i) est le rang de la matrice (a_{ij})).
On peut être un peu plus précis :

<u>Lemme 2</u>. Si D est algébrique sur K, $n \geq 2$ et $A \in M_n(D)$, alors det A admet un représentant de la forme $p^{-1}q$ où q est une expression polynômiale en les éléments de A et p une expression polynômiale en les éléments des $n-1$ premières colonnes de A.

<u>Preuve</u>. Par induction : soit $n=2$ et $A = \begin{pmatrix} a & b \\ c & d \end{pmatrix}$.
Si $ac = ca$, il n'y a rien à prouver.
Si $ac \neq ca$, le lemme 1 donne

$$ac.a^{-1} = p^{-1}q$$

où p ne dépend que de a et de c, et

$$ad - aca^{-1}b = p^{-1}(pad-qb) .$$

Soit $n > 2$. On peut supposer que $\det A \neq \overline{o}$.
Des propriétés de ce déterminant (voir références), on peut aussi supposer, quitte à multiplier det A par $\overline{-1}$, que $A_{11} \neq 0$.
Pour $i \geq 2$, on soustrait la première ligne multipliée à gauche par $A_{i1}A_{11}^{-1}$ de la $i^{\text{ème}}$.

Par le lemme 1, on écrit

$$A_{i1} A_{11}^{-1} = p_{(i)}(A_{11},A_{i1},A_{i1})^{-1} \; q_{(i)}(A_{11},A_{i1}) \; .$$

On remarque alors que la matrice

$$\begin{pmatrix} A_1 \\ p_{(2)}A_2 - q_{(2)}A_1 \\ . \\ . \\ . \\ p_{(n)}A_n - q_{(n)}A_1 \end{pmatrix}$$

est de la forme

$$\begin{pmatrix} A_{11} & B \\ 0 & \\ . & \\ . & C \\ 0 & \end{pmatrix}$$

et que les A_{in} n'apparaissent pas dans les $n-2$ premières colonnes de C . On applique l'hypothèse d'induction . □

Des lemms 1 et 2, on déduit le

<u>Corollaire</u>. Pour tout couple d'entiers $(\ell,n \geq 2)$, il existe un entier s des polynômes $Q_1,\ldots,Q_s$ en les variables non commutatives X_{ij} (où $1 \leq i,j \leq n$) et des polynômes $P_1,\ldots,P_s$ en les $X_{i\lambda}$ (où $1 \leq i \leq n$ et $1 \leq \lambda \leq n-1$) tels que si $[D:K] \leq \ell$, alors pour toute matrice $A \in M_n(D)$, il existe $\delta \in \{1,\ldots,s\}$ pour lequel

$$P_\delta(1_{11},\ldots,A_{1,n-1} \; , \; A_{21},\ldots,A_{2,n-1},\ldots,A_{n,n-1}) \neq 0$$

et

$$P_\delta(A_{11},\ldots,A_{n,n-1})^{-1} \; Q_\delta(A_{11},\ldots,A_{nn})$$

soit un représentant de det A.

Pour n=1, posons $P_1=1$ et $Q_1=X_{11}$.
<u>L'élimination linéaire pour la classe des corps de rang</u> $\leq \ell$ <u>sur leur centre s'exprime alors à l'aide de ces numérateurs et dénominateurs</u> Q_δ <u>et</u> P_δ<u>-ci de représentants de déterminants de DIEUDONNE</u> :
La formule Ψ_{mn} est équivalente sur tous les corps de cette classe à une disjonction de conjonctions de formules de la forme

$P_\delta \neq 0$, $Q_\delta \neq 0$ et $Q_\delta = 0$.

Revenons à l'élimination linéaire -2 sur les corps de rang 4 sur leur centre.

Soit $A = \begin{pmatrix} a & b \\ c & d \end{pmatrix}$ avec $ac \neq ca$.

Ecrivons $ca^{-1} = p^{-1}q$ où p et q sont déterminés par la preuve du lemme 1.

Dès lors

$$\det A = \overline{ap^{-1}(pd-qb)} \quad ,$$

et on vérifie que

$$ap^{-1}(pd-qb) = (a(ac-ca)^{-1}Q_2$$

où Q_2 est le premier membre de ([9]) . C'est le représentant de det A fourni par le corollaire.

D'autre part, on vérifie que

$$\begin{aligned} \det A &= \overline{c^{-1}} \det \begin{pmatrix} ca & cb \\ c & d \end{pmatrix} \\ &= c^{-1} \det \begin{pmatrix} \frac{(ca-ac)^2}{c} & \frac{(ca-ac)(cb-ad)}{d} \end{pmatrix} \\ &= ((ac-ca)c)^{-1} Q_1 \end{aligned}$$

où Q_1 est le premier membre de ([8]) . <u>On peut donc relier</u> Q_1 <u>et</u> Q_2 :

$$M_1 = ((ac-ca)c)^{-1}Q_1 \quad \text{et} \quad M_2 = (a(ac-ca)^{-1}Q_2$$

sont des expressions rationnelles en a, b, c et d qui, lorsqu'elles sont définies, diffèrent multiplicativement par un produit de commutateurs sur tous les corps de rang 4 sur leur centre (cette relation d'équivalence entre expressions rationnelles a un sens pour toute classe de corps. Elle est compatible avec la multiplication et donne un groupe commutatif par passage du quotient).

On peut vérifier cela directement : on calcule grâce à l'identité de WAGNER que l'égalité $aQ_1 = cQ_2$ est une identité polynômiale sur tous les corps de rang 4 sur leur centre. On en déduit alors que

$$M_1 = c^{-1}(a(ac-ca))^{-1} c(a(ac-ca))M_2$$

est une identité rationnelle sur tous ces corps.

Comme on peut s'en douter, ce texte doit beaucoup à des discussions de son auteur avec les trois premiers signataires de [8].

Bibliographie

[1] E. ARTIN, Geometric algebra, Interscience pub. New-York, 1957.

[2] J. BARWISE (ed.), Handbook of Mathematical Logic, North Holland (1977).

[3] Ch. BERLINE, rings which admit elimination of quantifiers, J. Symbolic Logic 46 (1980), 56-58.

[4] M. BOFFA, A. MACINTYRE and F. POINT, The quantifier elimination problem for rings without nilpotent elements and for semi-simple rings, Proc. of the Conference on Application of Logic to Algebra and Arithmetic Held at Bierutowice-Karpacz, Lecture Notes in Maths n°834, Springer-Verlag (1980), 20.

[5] M. BOFFA, Elimination des quantificateurs en algèbre, Bull. Soc. Math. Belg. (série B), 32 (1980), p. 107-133.

[6] M. BOFFA, Linear elimination, notes non publiées (Oberwolfach, Janvier 1984).

[7] M. BOFFA, Linear elemination, résumés des exposés du Modelltheorie-Tagung. 1984, Math. Forschungsinstitut Oberwolfach.

[8] M. BOFFA, C. MICHAUX, F. POINT, P. VAN PRAAG, L'élimination linéaire dans les corps, C.R. Acad. Sci. Paris, T.300, série 1, n°11, 1985.

[9] J. DIEUDONNE, Les déterminants sur un corps non commutatif, Bull. Soc. Math. France, 71 (1943), 27-45.

[10] P.K. DRAXL, Skew Fields, Cambridge University Press, 1982.

[11] A. LEROY, Les octaves de CAYLEY ont l'élimination linéaire -2 (à paraître).

[12] A. MACINTYRE, on ω_1-categorical theories of fields, Fund. Math. 71 (1971), 1-25.

[13] J.-F. PABION, Logique mathématique, Hermann, 1976.

[14] F. POINT, Sur l'élimination linéaire, C.R. Acad. Sc. Paris, 295, série 1, 1982, p. 211-213.

[15] F. POINT, Quantifier elimination for projectable L-groups and Linear elimination for rings, thèse, Mons, 1983.

[16] L. VAN DEN DRIES, Model theory of fields, thèse, Utrecht, 1978.

[17] P. VAN PRAAG, L'élimination linéaire dans les corps et le déterminant de DIEUDONNE, publications de l'Institut de Mathématique pure et appliquée de l'Université Catholique de Louvain rapport n°66, avril 1985.

[18] P. VAN PRAAG, Remarques sur l'élimination linéaire dans les corps et le déterminant de DIEUDONNE, Bull. Soc. Math. de Belg. série A, numéro spécial d'hommage à Guy HIRSCH, à paraître.

COMMUTANTS DES MODULES DE TYPE FINI

SUR LES ALGEBRES NOETHERIENNES

Gérard Cauchon

U.E.R. des Sciences de Reims
Département de Mathématiques
B.P. 347
51062 REIMS CEDEX (FRANCE)

Dans tout ce qui suit, k désigne un corps commutatif et A une k-algèbre associative et unitaire.

On dit que A vérifie le théorème de Nullstellensatz à gauche si, pour tout A-module à gauche simple S, on a :

$$\dim_k(\mathrm{End}_A(S)) < +\infty.$$

I. LES THEOREMES DE QUILLEN ET GABRIEL

<u>Définition</u> : Soit R un anneau commutatif, unitaire, intègre et B une R-algèbre associative unitaire. On dit que B vérifie <u>la propriété de platitude générique à gauche</u> si, pour tout B-module à gauche de type fini M, il existe un élément non nul c de R tel que M_c soit un R_c-module libre (où M_c et R_c désignent les localisés respectifs de M et R par rapport au système multiplicatif des puissances de c). On sait (voir par exemple [7]) qu'il en est ainsi lorsque B est une R-algèbre de type fini avec un système générateur $x_1,\ldots,x_p$ vérifiant :

$$[x_i, x_j] \in R + R\,x_1 + \ldots + R\,x_p$$

quels que soient $i,j \in \{1,\ldots,p\}$.

On en déduit immédiatement :

<u>Proposition 1</u> : Si A est une k-algèbre de Weyl ou si A est l'algèbre enveloppante d'une k-algèbre de Lie de dimension finie, alors l'anneau A[t] des polynômes en une indéterminée t, à coefficients dans A, vérifie la propriété de platitude générique à gauche (et aussi à droite) comme k[t]-algèbre.

Le résultat suivant, dû à Irving [4] permet de construire d'autres exemples pour lesquels ce phénomène se produit :

Proposition 2 : Si A se déduit de k par une succession d'extensions de Ore, alors, l'anneau de polynômes $A[t]$ vérifie la propriété de platitude générique comme $k[t]$ - algèbre.

On en déduit :

Proposition 3 : Si $A = k[G]$ où G est un groupe fini-par-polycyclique, alors, l'anneau de polynômes $A[t]$ vérifie la propriété de platitude générique comme $k[t]$-algèbre.

L'importance de la propriété de platitude générique a été soulignée par Quillen qui a démontré en 1969 le résultat suivant:

Théorème 1 [6] : Si A est l'algèbre enveloppante d'une k-algèbre de Lie de dimension finie et si S est un A-module à gauche simple, alors $\mathrm{End}_A(S)$ est algébrique sur k (identifié aux endomorphismes scalaires de S).

En fait, la démonstration de ce théorème fournit encore le résultat suivant :

Proposition 4 : Supposons que l'anneau de polynômes $A[t]$ vérifie la propriété de platitude générique à gauche comme $k[t]$-algèbre. Soit M un A-module à gauche de type fini et Ω un sous-corps (non nécessairement commutatif) de $\mathrm{End}_A(M)$ qui contient k. Alors Ω est algébrique sur k.

Démonstration : Supposons qu'il existe un élément α de Ω qui soit transcendant sur k. On munit M d'une structure de $A[t]$-module à gauche en posant $tm=\alpha(m)$ pour tout $m \in M$. Soit $c \in k[t] \setminus \{0\}$ tel que M_c soit un $k[t]_c$-module libre. Puisque α est transcendant sur k, pour tout polynôme non nul $f = \varepsilon_o + \varepsilon_1 t + \ldots + \varepsilon_n t^n \in k[t]$ $(\varepsilon_i \in k)$, l'homothétie de rapport f qui n'est autre que $\varepsilon_o \mathrm{Id}_M + \varepsilon_1\alpha + \ldots + \varepsilon_n\alpha^n$ est un élément non nul de Ω, donc un automorphisme du A-module à gauche M. Appliquant ce résultat à $f=c$, on en déduit que $M_c = M$.

Soit alors $\mathcal{B} = (b_i)_{i \in I}$ une base du $k[t]_c$-module libre de M, $i_o \in I$ et $f \in k[t] \setminus \{0\}$. Soit m l'antécédent de b_{i_o} par l'homothétie de rapport f de M. Notons $m = \sum_{i \in I} u_i b_i$ sa décomposition dans la base B. (Les u_i sont des éléments presque tous nuls de $k[t]_c$). Alors, multipliant par f, il vient $b_{i_o} = \sum_{i \in I} f u_i b_i$. On en déduit que $f u_{i_o} = 1$, donc que $\frac{1}{f} = u_{i_o} \in k[t]_c$.

Il résulte de ceci que $k[t]_c = k(t)$ le corps des fractions rationnelles en l'indéterminée t, ce qui est impossible. Et, nécessairement, Ω est algébrique sur k.

De la proposition 4, on déduit :

Proposition 5 :

Supposons que l'anneau de polynômes $A[t]$ vérifie la propriété de platitude générique comme $k[t]$-algèbre. Alors, si M est un A-module à gauche de longueur finie, $\mathrm{End}_A(M)$ est algébrique sur k.

Démonstration : Par récurrence sur la longueur ℓ de M. Si $\ell = 1$, $\mathrm{End}_A(M)$ est un corps et il suffit d'appliquer la proposition 4. Supposons $\ell > 1$ et le résultat vrai pour tout A-module à gauche M' de longueur finie $\ell' < \ell$. Soit $\alpha \in \mathrm{End}_A(M)$. Supposons que, pour tout polynôme non nul $P(t) = \varepsilon_o + \varepsilon_1 t + \ldots + \varepsilon_n t^n$ de $k[t]$ $(\varepsilon_i \in k)$, l'endomorphisme $P(\alpha) = \varepsilon_o \, \mathrm{Id}_M + \varepsilon_1 \alpha + \ldots + \varepsilon_n \alpha^n$ soit un automorphisme de M. Alors, l'ensemble Ω tous les endomorphismes de M de la forme $P(\alpha)^{-1} Q(\alpha)$ $(P \in k[t] \setminus \{0\}, Q \in k[t])$ serait un sous-corps de $\mathrm{End}_A(M)$ contenant k et transcendant sur k, ce qui contredit la proposition 4. Il existe donc $P \in k[t] \setminus \{0\}$ tel que l'endomorphisme $\alpha' = P(\alpha)$ ne soit pas bijectif. Alors $M' = \alpha'(M)$ est de longueur finie $\ell' < \ell$. La restriction de α' à M' est un élément de $\mathrm{End}_A(M')$, et donc est algébrique sur k par l'hypothèse de récurrence. Il existe donc $\omega_o, \ldots, \omega_s \in k$ avec $\omega_s \neq 0$, tels que

$$\omega_o m' + \omega_1 \alpha'(m') + \ldots + \omega_s \alpha'^s(m') = 0 \qquad (\forall\, m' \in M').$$

Il en résulte que $\omega_o \alpha' + \omega_1 \alpha'^2 + \ldots + \omega_s \alpha'^{s+1} = 0$, de sorte que α annule le polynôme $T(t) = \omega_o P(t) + \omega_1 P(t)^2 + \ldots + \omega_s P(t)^{s+1} \in k[t] \setminus \{0\}$.

La démonstration du théorème de Nullstellensatz pour les algèbres enveloppantes d'algèbres de Lie de dimension finie ([3] 2-6-9 p.90) se fait au moyen du résultat suivant, dû a Gabriel :

Théorème 2 :

Faisons les hypothèses suivantes :

(H_1) Le corps k est parfait

(H_2) Pour tout corps commutatif k', extension algébrique de k, l'anneau $A' = A \otimes_k k'$ est noethérien à gauche.

Alors, si S est un A-module à gauche semi-simple de longueur finie, et si k'

est une extension algébrique de k, $S' = S \otimes_k k'$ est un A'-module à gauche, semi-simple de longueur finie.

Démonstration : S' est un A'-module de type fini. Il suffit donc de montrer que S' est un A'-module à gauche semi-simple. Pour ceci, on suppose d'abord que k' est une extension finie de k et on recopie (en changeant les notations) la démonstration de ([3] 1.2.19 c) p. 15). Dans le cas général, on recopie la démonstration de ([3] 2.6.9. a)b) p.90-91).

On en déduit, comme en ([3] 2.6.9, c):

Théorème 3 :

Faisons les hypothèses suivantes :

(H_1) Le corps k est parfait

(H_2) Pour tout corps commutatif k' enxtension algébrique de k, l'anneau $A' = A \otimes_k k'$ est noethérien à gauche.

(H_3) pour tout corps commutatif k' extension algébrique de k, et tout $A' = A \otimes_k k'$-module à gauche simple S', $\mathrm{End}_{A'}(S')$ est algébrique sur k'.

Alors, A vérifie le théorème de Nullstellensatz à gauche.

Démonstration : Prenons pour k' la cloture algébrique de k et posons $A' = A \otimes_k k'$. On observe que, par (H_3), si S' est un A'-module à gauche simple, on a $\mathrm{End}_{A'}(S') = k'$. Par suite, si Σ est A'-module à gauche semi-simple de longueur finie, $\mathrm{End}_{A'}(\Sigma)$ est isomorphe à un produit fini d'anneaux de matrices carrées à coefficients dans k', donc est de dimension finie sur k'. Considérons un A-module à gauche simple S, et posons $\Omega = \mathrm{End}_A(S)$. Alors, on a $\Omega \otimes_k k' = \mathrm{End}_{A'}(S \otimes_k k')$. Par le théorème 2, $S \otimes k'$ est un A'-module à gauche semi-simple de longueur finie. Donc $\Omega \otimes_k k'$ est de dimension finie sur k' et, par suite, Ω est de dimension finie sur k.

Si A est l'algèbre enveloppante d'une algèbre de Lie g de dimension finie sur k (resp. si A est une algèbre de Weyl d'indice n sur k) et si k' est un corps commutatif extension de k, alors $A' = A \otimes_k k'$ est l'algèbre enveloppante de l'algèbre de Lie $g' = g \otimes_k k'$ qui est dimension finie sur k' (resp. $A' = A \otimes_k k'$ est l'algèbre de Weyl d'indice n sur k').

Si G est un groupe, si $A = k[G]$ et si k' est un corps commutatif extension de k, alors $A' = A \otimes_k k' = k'[G]$. Il résulte alors des propositions 1, 3 et 4:

Proposition 6 :

La k-algèbre A vérifie les hypothèses (H_2) et (H_3) du théorème 3 dans les trois cas suivants :

- A est l'algèbre enveloppante d'une algèbre de Lie de dimension finie sur k.
- A est une algèbre de Weyl sur k
- A = k[G] où G est un groupe fini-par-polycyclique.

Le théorème 3 et cette proposition nous redonnent les résultats connus suivants :

Corollaire 1 [3] :

Si le corps k est de caractéristique nulle et si A est l'algèbre enveloppante d'une algèbre de Lie de dimension finie sur k ou bien une algèbre de Weyl sur k, alors A vérifie le théorème du Nullstenllensatz à gauche.

Corollaire 2 [5] :

Si le corps k est parfait et si A = k[G] où G est un groupe fini par polycyclique, alors A vérifie le théorème de Nullstellensatz à gauche.

Si A est comme dans le corollaire 1 et si le corps k est de caractéristique non nulle, alors A est une algèbre affine à identité polynômiale. Donc, par un théorème de Amitsur et Procesi [1], A vérifie encore le théorème du Nullstellensatz à gauche. Cela suggère la question suivante :

Question 1 : Le théorème 3 (et en particulier le corollaire 2) est-il encore vrai lorsqu'on ne fait plus l'hypothèse "k est parfait"?

Remarque : Si le théorème 3 est encore vrai sans l'hypothèse (H_1), sa démonstration ne peut plus utiliser le théorème 2 de Gabriel puisque, dans ce théorème l'hypothèse (H_1) est indispensable. (Si k n'est pas parfait et si A est un corps commutatif extension finie non séparable de k, alors A est un A-module simple qui contre-exemple le théorème 2).

II- LE THEOREME DE SMALL

Dans une communication au congrès d'Anvers de 1983, Small a démontré que le théorème 3 est encore vrai si on supprime l'hypothèse (H_1), à condition de renforcer l'hypothèse (H_3) et a fourni une réponse complète à la question 1 en ce qui concerne le corollaire 2.

Théorème 1 [8] : Faisons les hypothèses suivantes :

(H_2) Pour tout corps commutatif k' extension algébrique de k, l'anneau $A' \otimes_k k'$ est noethérien à gauche.

(H_3') Pour tout corps commutatif k' extension algébrique de k, et tout $A' = A \otimes_k k'$-module à gauche de longueur finie M', $\mathrm{End}_{A'}(M')$ est algébrique sur k'. Alors, A vérifie le théorème de Nullstellensatz à gauche.

Remarque : Pour mieux comparer les hypothèses (H_3) et (H_3'), on peut observer que :

a) (H_3') équivaut à :

Pour tout corps commutatif k' extension algébrique de k et tout $A' = A \otimes_k k'$-module à gauche simple S', le corps $\Omega' = \mathrm{End}_{A'}(S')$ est algébrique sur k', ainsi que tous les anneaux de matrices carrées à coefficients dans Ω'.

b) On ne connait pas d'exemple de corps gauche Ω' algébrique sur un sous-corps central k' et tel qu'il existe une matrice carrée à coefficients dans Ω' qui ne soit pas algébrique sur k'.

Démonstration du Théorème 1 : Soit k' la clôture algébrique de k et S un A-module à gauche simple. On pose $A' = A \otimes_k k'$ et $S' = S \otimes_k k'$. Pour montrer que le corps $\Omega = \mathrm{End}_A(S)$ est de dimension finie sur k, il suffit, comme dans la démonstration du théorème 3 du paragraphe I, de montrer que $\Omega' = \mathrm{End}_{A'}(S')$ est de dimension finie sur k'. Pour ceci, on fait deux observations.

a) Soit k'' un corps intermédiaire entre k et k', extension finie de k. Posons $A'' = A \otimes_k k''$ et $S'' = S \otimes_k k''$. S'' est un A-module à gauche semi-simple de longueur finie, donc un A''-module à gauche de longueur finie. Alors, pour tout corps k'' intermédiaire entre k et k', et extension finie de k, $\Omega'' = \Omega \otimes_k k''$ est algébrique sur k''. Il en résulte que $\Omega' = \Omega \otimes_k k'$ est algébrique sur k'.

b) Par ([1] p. 88), le treillis des idéaux à droite de Ω' est isomorphe au treillis des sous A-modules de S'. Donc, par (H_2), Ω' est noethérien à droite. Soit N le radical premier de Ω'. Ω'/N est un anneau semi-premier, noethérien à droite et algébrique sur k'. Tous ses éléments réguliers sont donc inversibles. Par suite, $\frac{\Omega'}{N}$ est un anneau semi-simple et, k' étant algébriquement clos, $\frac{\Omega'}{N}$ est un produit fini d'anneaux de matrices carrées sur k'. Donc $\dim_{k'}(\frac{\Omega'}{N}) < +\infty$.

Comme N est nilpotent, on en déduit que $\dim_{k'}(\Omega') < +\infty$, ce qui achève la démonstration.

Corollaire : Si $A = k[G]$ où G est un groupe fini-par-polycyclique, alors A vérifie le théorème du Nullstellensatz à gauche.

Démonstration : On sait déjà que A vérifie (H_2) (I prop. 6). Pour montrer qu'elle vérifie (H_3'), considérons un corps commutatif k' extension algébrique de k, et posons $A' = A \otimes_k k' = k'[G]$. Alors, par les propositions 3 et 5 du paragraphe I appliquées à la k'-algèbre A', pour tout A'-module à gauche M' de longueur finie, $\mathrm{End}_{A'}(M')$ est algébrique sur k' ; ce qu'il fallait démontrer.

III - SUPPRESSION DE L'HYPOTHESE (H_1) ET ALLEGEMENT DE L'HYPOTHESE (H_3) dans LE THEOREME 3 DU PARAGRAPHE I.

Dans le théorème de Small, l'hypothèse (H_2) est indispensable. En effet, si A est un corps commutatif extension algébrique infinie de k, il est clair que A vérifie l'hypothèse (H_3'), mais ne vérifie par le théorème du Nullstellensatz à gauche. En fait, nous allons voir que l'hypothèse (H_2) à elle seule, a des conséquences remarquables :

Lemme 1 : Soit M un A-module à gauche de type fini et k' un sous-corps commutatif de $\mathrm{End}_A(M)$ qui contient k. Supposons que l'anneau $A' = A \otimes_k k'$ soit noethérien à gauche. Alors, k' est une extension de type fini de k.

Démonstration : Considérons M comme un A-k'-bimodule. Alors $M \otimes_k k'$ est naturellement muni d'une structure de $A \otimes_k k'$-$k' \otimes_k k'$-bimodule. Comme k' est un corps, M est un k'-module libre à droite et, si $B = (b_i)_{i \in I}$ est une base de M comme k'-module, $B' = (b_i \otimes 1)_{i \in I}$ est une base de $M \otimes_k k'$ comme $k' \otimes k'$-module. $M \otimes_k k'$ est un $A \otimes_k k'$-module à gauche de type fini, donc noethérien, par hypothèse. Comme $M \otimes_k k'$ est un $k' \otimes_k k'$-module libre, on voit facilement que l'application $J \to (M \otimes_k k')J$ est injective croissante du treillis des idéaux de $k' \otimes_k k'$ dans le treillis des sous-modules de $M \otimes_k k'$ considéré comme un $A \otimes_k k'$-module à gauche. Par suite, $k' \otimes_k k'$ est un anneau noethérien. L'application qui, à tout corps L intermédiaire entre k et k' fait correspondre le noyau du morphisme canonique :

$$k' \otimes_k k' \longrightarrow k' \otimes_L k'$$

est strictement croissante du treillis des corps intermédiaires entre k et k' dans le treillis des idéaux de $k' \otimes_k k'$. Donc les corps intermédiaires entre k et k' vérifient la condition de chaine ascendante finie, et k' est bien une extension de type fini de k.

<u>Proposition 1</u> :
Supposons que A vérifie l'hypothèse (H_2) (I Th. 3).
Soit M un A-module à gauche de type fini et Ω un sous-corps (non nécessairement commutatif) de $End_A(M)$. Alors, si Ω est algébrique sur k, il est de dimension finie sur k.
<u>Démonstration</u> : Soit k' un sous-corps commutatif maximal de Ω. Par (H_2), l'anneau $A' = A \otimes_k k'$ est noethérien à gauche. Donc par le lemme 1, $[k':k] < +\infty$.
Notons Z le centre de Ω. On a $k' \supset Z \supset k \Rightarrow [k':Z] < +\infty \Rightarrow [\Omega:Z] = [k':Z]^2 < +\infty$ ([2] p. 112, Th. 2) $\Rightarrow [\Omega:k] = [\Omega:Z] \times [Z:k] < +\infty$.

On en déduit évidemment le résultat suivant qui généralise le théorème 3 du paragraphe I et le théorème 1 du paragraphe II :
<u>Théorème 1</u> :
Faisons les hypothèses suivantes :
(H_2) Pour tout corps commutatif k' extension algébrique de k, l'anneau $A' = A \otimes_k k'$ est noethérien à gauche.
(H_3') Pour tout A-module à gauche simple S, $End_A(S)$ est algébrique sur k.
Alors, A vérifie le théorème du Nullstellensatz à gauche.

On en déduit également, par les propositions 1,2,3,4 du paragraphe I:
<u>Théorème 2</u> :
Supposons que A soit l'algèbre enveloppante d'une k-algèbre de Lie de dimension finie, ou que A soit une k-algèbre de Weyl, ou que A soit de la forme $k[G]$ où G désigne un groupe fini-par-polycyclique, ou bien encore que A se déduise de k par une succession finie d'extension de Ore. Alors A possède la propriété suivante :
Si M est un A-module à gauche de type fini, et si Ω est un sous-corps de $End_A(M)$ qui contient k, alors Ω est de dimension finie sur k.

Terminons par la question suivante qui est une généralisation naturelle du problème de Kurosh :
<u>Question</u> : Le théorème 1 et la proposition 1 du paragraphe III sont-ils encore vrais lorsqu'on remplace l'hypothèse (H_2) par :

"La k-algèbre A est affine et noethérienne à gauche" ?

REFERENCES

[1] Amitsur et Procesi : Annali de Mathematica, 71, (1966), p.61-71.

[2] Bourbaki : Algèbre. Chap. VIII.

[3] Dixmier : Algèbres enveloppantes. Gauthier-Villars. Paris 1974.

[4] Irving : Generic Flatness and the nullstellensatz for Ore extensions. Comm. in Algebra. Vol.7, N°3,1979,p.259-278.

[5] Lorenz : Primitive Ideals of group algebras of supersoluble groups. Math. Ann. 225, 1977, p.115-122.

[6] Quillen : On the endomorphism ring of a simple module over an enveloping algebra. Proc. Am. Math. Soc., 21, 1969, p.171-172.

[7] Renault : Algèbre non commutative. Gauthier-Villars. Paris 1975.

[8] Small : F. Van Oystaeyen (ed.) Methods in Ring Theory. 1984.

DEUX APPLICATIONS DU PRINCIPE D'ADDITIVITE

A. Hudry

1. INTRODUCTION.

A. Joseph et L.W. Small [5] ont démontré que si A et B sont certaines k-algèbres de type fini noethériennes apparaissant dans l'étude des algèbres enveloppantes des k-algèbres de Lie de dimensions finies, avec A sous-k-algèbre de B et B de type fini en tant que A-module à droite et à gauche, alors pour tout idéal premier P de B, le rang de Goldie de $B/_P$ vérifie le principe d'additivité suivant :

$$\text{rk } B/P = \sum_{i=1}^{n} \mu_i \text{ rk } A/p_i$$

où $\mu_1,\ldots,\mu_n$ sont des entiers naturels strictement positifs et où $\{p_1,p_2,\ldots,p_n\}$ est l'ensemble des idéaux premiers de A minimaux au-dessus de $A \cap P$. W. Borho [2], R.B. Warfield [14] et J.T. Stafford [13] ont généralisé ce résultat. Dans [4] nous avons donné un principe d'additivité du rang et du rang réduit de Goldie pour des bimodules qui généralise tous les principes d'additivité déjà connus. Ici il va être présenté deux applications de ce principe d'additivité. La première concerne les actions de groupes : étant donnés un anneau unitaire A et un groupe fini G d'automorphismes de A tel que $|G|^{-1} \in A$, il est établi que pour toute une classe d'idéaux de Small I de A, le rang (resp. le rang réduit) de $A/_I$ est lié aux co-rangs des idéaux premiers de A^G minimaux au-dessus de $I \cap A^G$ par un principe d'additivité ; ce résultat généralise strictement celui obtenu pour les idéaux premiers de Goldie de A, dans [8] par M. Lorenz, S. Montgomery et L.W. Small. La deuxième application concerne les extensions triangulaires de B. Lemonnier [6] : une généralisation d'un théorème de R.B. Warfield [14] est obtenue.

2. TERMINOLOGIE.

Tous les anneaux considérés ici sont unitaires et lorsqu'un anneau est dit noethérien, il est entendu qu'il l'est à droite et à gauche. Lorsque M désigne un (B,A)-bimodule, il est entendu qu'il s'agit en particulier d'un B-module à gauche et d'un A-module à droite ; l'annulateur de M considéré comme B-module (resp. A-module) à gauche (resp. à droite) est noté $\ell_B(M)$ (resp. $r_A(M)$). Par idéal I d'un anneau B, on entend un idéal bilatère ; la partie multiplicative des éléments de B réguliers modulo I est alors notée $\mathscr{C}_B(I)$; I est dit idéal de Small de B si $B/_I$ admet un anneau classique de fractions (à droite et à gauche) artinien. Soit $\mathcal{T}_B$ la sous-catégorie localisante de B-Mod, qui est le noyau du foncteur $\text{Hom}_B(. , E(B/\text{rad } B))$ où rad B désigne le radical premier de B et où E(B/rad B) est une enveloppe injective du

B-module à gauche B/rad B. D'après [1], un B-module à gauche M a un rang réduit fini si et seulement si son image $\ell_B(M)$ dans la catégorie quotient $\mathcal{L}_B = B\text{-Mod}/\mathcal{T}_B$ est de longueur finie ; on pose alors $\text{rrk}({}_BM) = \text{long}\ \ell_B(M)$; si λ_B est l'ensemble topologisant et idempotent des idéaux à gauche de B associé à $\mathcal{T}_B$, cela signifie que le treillis des sous-modules λ_B-clos de M est de longueur finie. On rappelle qu'un B-module à gauche est de rang de Goldie fini si son enveloppe injective est somme directe d'un nombre fini d'injectifs indécomposables ; ce nombre, indépendant d'une telle décomposition, est noté $\text{rk}({}_BM)$. Pour tout ce qui concerne la dimension de Gelfand-Kirillov (GK-dimension), on peut se référer à [3], [5] et [7]. On rappelle ici les résultats obtenus dans [4] concernant le principe d'additivité pour les bimodules.

Proposition 1. — *Soit* M *un* (B,A)*-bimodule pour lequel les propriétés suivantes sont vérifiées :*

a) M *est sans torsion et fidèle comme* A*-module à droite,*

b) A *a un anneau classique de fractions à droite artinien à droite,*

c) M *considéré comme* B*-module à gauche a un rang réduit fini et* M *est sans* λ_B*-torsion.*

Alors ${}_BM$ *a un rang de Goldie fini et on a :*

$$\text{rk}({}_BM) = \sum_{i=1}^{n} \mu_i \ \text{rk}\ A/p_i \quad \textit{et} \quad \text{rrk}({}_BM) = \sum_{i=1}^{n} \nu_i \ \text{rk}\ A/p_i$$

où les μ_i *et les* ν_i *pour* $i \in \{1,2,\ldots,n\}$ *sont des entiers naturels strictement positifs tels que* $\mu_i \leq \nu_i$ *et où* $\{p_1,p_2,\ldots,p_n\} = \text{Spec}_{\min} A$.

Proposition 2. — *Soit* M *un* (B,A)*-bimodule pour lequel les propriétés suivantes sont vérifiées :*

a) *Tout anneau quotient premier de* A *est de Goldie à droite,*

b) M *considéré comme* B*-module à gauche a un rang réduit fini.*

Alors il existe un ensemble fini $\{p_1,\ldots,p_n\}$ *d'idéaux premiers de* A *contenant l'ensemble des idéaux premiers de* A *minimaux au-dessus de* $r_A(M/\lambda_B(M))$ *tel que l'on ait :*

$$\text{rrk}({}_BM) = \sum_{i=1}^{n} \nu_i \ \text{rk}\ A/p_i$$

où les ν_i *pour* $i \in \{1,\ldots,n\}$ *sont des entiers naturels strictement positifs.*

(Les idéaux premiers p_i de la proposition 2 sont les annulateurs dans A des quotients successifs d'une chaîne de sous-(B,A)-bimodules M_j de M de la forme :

$$M_0 = \lambda_B(M) \subsetneq M_1 \subsetneq \ldots \subsetneq M_{j-1} \subsetneq M_j \subsetneq \ldots \subsetneq M_r = M$$

avec M_j minimal dans l'ensemble des sous-(B,A)-bimodules λ_B-clos de M contenant strictement M_{j-1} ; si les idéaux premiers p_i sont de Goldie à droite, l'hypothèse a) de la proposition 2, peut être omise.)

3. ACTIONS DE GROUPES ET PRINCIPE D'ADDITIVITE.

Dans cette section et la suivante, A désigne un anneau unitaire et G un groupe fini d'automorphismes de A tel que $|G|^{-1} \in A$. L'anneau de groupe gauche ("skew group ring") $A * G$ associé à A et G est le produit croisé de G sur A dont la multiplication est définie par : $(rx)(sy) = r\, s^{x^{-1}} xy$ pour tout $r,s \in A$ et tout $x,y \in G$. $A * G$ est alors une extension normalisante libre de A de base G et $e = |G|^{-1} \sum_{x \in G} x$ est un idempotent de $A * G$ vérifiant pour tout $x \in G$, $ex = xe = e$ et pour tout $r \in A^G$, $er = re$, de plus on a :

$$e(A * G)e = e\, A^G\, e = e\, A^G \simeq A^G .$$

Un idéal I de A est dit G-invariant si $I^x = I$ pour tout $x \in G$. Tout idéal I de A contient un plus grand idéal G-invariant à savoir $\tilde{I} = \bigcap_{x \in G} I^x$. G agit sur Spec A de sorte que pour tout idéal premier P de A l'idéal $\tilde{P}$ est un idéal semi-premier G-invariant. Deux idéaux premiers P_1 et P_2 de A sont dits G-incomparables si les idéaux semi-premiers $\tilde{P}_1$ et $\tilde{P}_2$ sont incomparables, c'est-à-dire si tout idéal premier de la G-orbite de P_1 est incomparable avec tout idéal premier de la G-orbite de P_2. Les idéaux premiers P_1 et P_2 de A vérifient la condition $\tilde{P}_1 = \tilde{P}_2$ si et seulement si la G-orbite de P_1 coïncide avec la G-orbite de P_2. Pour un idéal de Small I de A, on considère alors la condition naturelle suivante :

(*) : "Deux idéaux premiers quelconques de A minimaux au-dessus de I et non dans la même G-orbite sont G-incomparables".

UNE CLASSE D'IDEAUX DE SMALL.

Pour un idéal I de Small de A, les cas suivants vont en particulier être envisagés.

Cas 1. — I est un idéal premier de Goldie (c'est le seul cas considéré par M. Lorenz, S. Montgomery et L.W. Small dans [8]).

Cas 2. — I est un idéal semi-premier de Goldie vérifiant la condition (*).

Cas 3. — A est noethérien (ou de manière moins restrictive de dimension de Krull à droite et à gauche définie) et I est un idéal de Small vérifiant la condition (*).

Cas 4. — A est une k-algèbre noethérienne de GK-dimension finie et I est un idéal GK-homogène dans A.

(Exemple : tout idéal I primaire (resp. classiquement primaire) de l'algèbre enveloppante d'une super k-algèbre de Lie (resp. d'une algèbre de Lie) de dimension finie).

Cas 5. — A est un anneau noethérien Krull-symétrique et I est un idéal Krull-homogène dans A.

(Exemple : tout idéal primaire (resp. classiquement primaire) d'un anneau noethérien à I.P. ou plus généralement totalement borné).

<u>Cas 6</u>. — I est un idéal de Small G-invariant.

Remarques.

1. Les cas 4 et 5 sont des cas particuliers du cas 3.

2. Dans les 6 cas précédents, I est un idéal de Small de A vérifiant la condition (*) et la condition (**) : " $A/\widetilde{I}$ est de rang réduit fini". La conclusion du lemme suivant est encore vérifiée pour la classe des idéaux de Small satisfaisant aux conditions (*) et (**).

Lemme 3. — *Dans les six cas précédents* $\widetilde{I}$ *est un idéal de Small vérifiant la condition* $\mathscr{C}_A(\widetilde{I}) \subset \mathscr{C}_A(I)$.

Démonstration.

<u>Cas 1</u>. Comme les anneaux $A/_I$ et $A/_{I^x}$ sont isomorphes pour tout $x \in G$, $\widetilde{I} = \bigcap_{x \in G} I^x$ est un idéal semi-premier de Goldie. Si X désigne un système de représentants des classes à droite dans G modulo le stabilisateur de I, on peut démontrer que $\{I^x | x \in X\}$ est l'ensemble des idéaux premiers de A minimaux au-dessus de $\widetilde{I}$ donc que l'on a :

$$\mathscr{C}_A(\widetilde{I}) = \bigcap_{x \in X} \mathscr{C}_A(I^x) = \bigcap_{x \in G} \mathscr{C}_A(I^x) \subset \mathscr{C}_A(I) .$$

<u>Cas 2</u>. Soit $\{P_1, P_2, \ldots, P_n\}$ l'ensemble des idéaux premiers de A minimaux au-dessus de l'idéal semi-premier de Goldie I. On suppose ici $n > 1$. Quitte à changer la numérotation des idéaux premiers minimaux au-dessus de I, on peut, compte tenu de l'hypothèse faite dans le cas 2, trouver une suite strictement croissante d'entiers naturels $(\nu_j)_{0 \leq j \leq k}$ avec $\nu_0 = 1$ et $\nu_k = n$ telle que

$$\widetilde{P}_{\nu_0} = \widetilde{P}_2 = \ldots = \widetilde{P}_{\nu_1}, \widetilde{P}_{\nu_1+1} = \ldots = \widetilde{P}_{\nu_2}, \ldots, \widetilde{P}_{\nu_{k-1}+1} = \ldots = \widetilde{P}_{\nu_k} ,$$

où les idéaux semi-premiers de Goldie $\widetilde{P}_{\nu_1}, \ldots, \widetilde{P}_{\nu_k}$ sont deux à deux incomparables. Comme dans le cas 1, on a pour $j \in \{1,2,\ldots,k\}$:

$$\mathscr{C}_A(\widetilde{P}_{\nu_j}) = \bigcap_{x_j \in X_j} \mathscr{C}_A(P_{\nu_j}^{x_j}) = \bigcap_{x \in G} \mathscr{C}_A(P_{\nu_j}^{x}) ,$$

où X_j désigne un système de représentants des classes à droite dans G, modulo le stabilisateur de P_{ν_j}. Comme on a $I = \bigcap_{i=1}^{n} P_i$, il vient :

$$\widetilde{I} = \bigcap_{i=1}^{n} \widetilde{P}_i = \bigcap_{j=1}^{k} \widetilde{P}_{\nu_j} = \bigcap_{j=1}^{k} \left(\bigcap_{x_j \in X_j} P_{\nu_j}^{x_j} \right).$$

Les idéaux premiers $P_{\nu_j}^{x_j}$ étant de Goldie, $\widetilde{I}$ est un idéal semi-premier de Goldie. L'ensemble des idéaux premiers de A minimaux au-dessus de $\widetilde{I}$ est $\bigcup_{j=1}^{k} \{P_{\nu_j}^{x_j} | x_j \in X_j\}$, ce qui implique :

$$\mathscr{C}_A(\tilde{I}) = \bigcap_{j=1}^{k}\Big(\bigcap_{x_j \in X_j} \mathscr{C}_A(P_{\nu_j}^{x_j})\Big) = \bigcap_{j=1}^{k}\Big(\bigcap_{x\in G} \mathscr{C}_A(P_{\nu_j}^{x})\Big) \subset \bigcap_{i=1}^{n} \mathscr{C}_A(P_i) = \mathscr{C}_A(I)$$

Cas 3. Soit $\{P_1,P_2,\ldots,P_n\}$ l'ensemble des idéaux premiers de A minimaux au-dessus de l'idéal de Small I. On applique alors à la racine $N = P_1 \cap \ldots \cap P_n$ de I les résultats du cas 2, et on obtient en conservant les notations précédentes :

$$\mathscr{C}_A(\tilde{N}) = \bigcap_{j=1}^{k}\Big(\bigcap_{x\in G} \mathscr{C}_A(P_{\nu_j}^{x})\Big) \subset \mathscr{C}_A(N) = \mathscr{C}_A(I) .$$

Pour tout $x \in G$, $\{P_i^x \mid i \in \{1,2,\ldots,n\}\}$ est l'ensemble des idéaux premiers de A minimaux au-dessus de l'idéal de Small I^x de A et la racine de I^x est $N^x = \bigcap_{i=1}^{n} P_i^x$, donc on a :

$$\mathscr{C}_A(N^x) = \bigcap_{i=1}^{n} \mathscr{C}_A(P_i^x) = \mathscr{C}_A(I^x) .$$

Pour tout $i \in \{1,2,\ldots,n\}$, il existe un unique entier $j \in \{1,2,\ldots,k\}$ tel que $i \in \{\nu_{j-1}+1,\ldots,\nu_j\}$; comme $\tilde{P}_i = \tilde{P}_{\nu_j}$, pour tout $x \in G$ il existe $y \in G$ tel que $P_i^x = P_{\nu_j}^y$, et par suite on a :

$$\mathscr{C}_A(\tilde{N}) \subset \mathscr{C}_A(P_{\nu_j}^y) = \mathscr{C}_A(P_i^x) .$$

Ceci implique pour tout $x \in G$:

$$\mathscr{C}_A(\tilde{N}) \subset \bigcap_{i=1}^{n} \mathscr{C}_A(P_i^x) = \mathscr{C}_A(I^x) .$$

Par suite on obtient :

$$\mathscr{C}_A(\tilde{N}) \subset \bigcap_{x\in G} \mathscr{C}_A(I^x) \subset \mathscr{C}_A(\tilde{I}) .$$

Comme $\tilde{N}$ est la racine de $\tilde{I}$, la condition de régularité de Small est satisfaite par $\tilde{I}$. Il en résulte que si $A/_{\tilde{I}}$ a un rang réduit fini (à droite et à gauche), alors $A/_{\tilde{I}}$ admet un anneau classique de fractions artinien, c'est-à-dire $\tilde{I}$ est un idéal de Small. On a alors :

$$\mathscr{C}_A(\tilde{N}) = \mathscr{C}_A(\tilde{I}) \subset \mathscr{C}_A(I) .$$

Cas 4. D'après [5], I est dans ce cas un idéal de Small et pour tout idéal premier P de A minimal au-dessus de I, on a :

$$\text{GK dim } A/_I = \text{GK dim } A/_P .$$

Soient P_1 et P_2 deux idéaux premiers de A minimaux au-dessus de I tels que $\tilde{P}_1 \neq \tilde{P}_2$. Si on avait par exemple $\tilde{P}_2 \subset \tilde{P}_1$ il existerait $x \in G$ tel que $P_2^x \subsetneq P_1$, et on obtiendrait la contradiction :

$$\text{GK dim } A/_I = \text{GK dim } A/_{P_1} < \text{GK dim } A/_{P_2^x} = \text{GK dim } A/_{P_2} = \text{GK dim } A/_I .$$

Donc P_1 et P_2 sont G-incomparables. La condition (*) est alors vérifiée par I et il suffit d'appliquer le cas 3.

Cas 5. La démonstration est analogue à celle du cas 4, en utilisant la K dim à la place de la GK dim.

Cas 6. Ce cas n'a été cité que pour mémoire.

LE PRINCIPE D'ADDITIVITE POUR DES IDEAUX DE SMALL.

Proposition 4. — *Soient* A *un anneau unitaire et* G *un groupe fini d'automorphismes de* A *tel que* $|G|^{-1} \in A$. *Pour tout idéal de Small* I *de* A *pour lequel* $\tilde{I} = \bigcap_{x \in G} I^x$ *est un idéal de Small vérifiant la condition* $\mathscr{C}_A(\tilde{I}) \subset \mathscr{C}_A(I)$ *(voir la liste des exemples détaillés dans lemme 3 précédent), les propriétés suivantes sont vérifiées :*

1) *Les idéaux premiers de* A^G *minimaux au-dessus de* $I \cap A^G$ *sont tous de Goldie et forment un ensemble fini* $\{p_1, p_2, \ldots, p_n\}$.

2) *Il existe* $\mu_1, \mu_2, \ldots, \mu_n, \nu_1, \nu_2, \ldots, \nu_n \in \mathbb{N}^*$ *avec pour tout* $i \in \{1,2,\ldots,n\}$ $\mu_i \leq \nu_i$ *tels que le rang de Goldie et le rang réduit (à gauche) de* $A/_I$ *vérifient :*

$$\mathrm{rk}\, A/_I = \sum_{i=1}^{n} \mu_i \,\mathrm{rk}\, A^G/_{p_i},$$

et

$$\mathrm{rrk}\, A/_I = \sum_{i=1}^{n} \nu_i \,\mathrm{rk}\, A^G/_{p_i}.$$

Remarque. — Cette proposition 4 appliquée au cas 1 du lemme précédent redonne le résultat principal obtenu par M. Lorenz, S. Montgomery et L.W. Small dans [8].

Démonstration. — Si N est la racine de I, la racine de $\tilde{I}$ est alors $\tilde{N}$. G agit sur $A/_{\tilde{I}}$ et comme $|G|^{-1} \in A$, on peut démontrer que l'on a :

$$(A/_{\tilde{I}})^G = A^G + \tilde{I}/_{\tilde{I}}.$$

Pour simplifier les écritures on va supposer $\tilde{I} = 0$. $\tilde{N}$ est alors le radical premier de A et il vérifie la condition : $\mathscr{C}_A(0) = \mathscr{C}_A(\tilde{N}) \subset \mathscr{C}_A(I)$. Comme $|G|^{-1} \in A$, le radical premier de A^G est :

$$\tilde{N} \cap A^G = (\tilde{N})^G.$$

Puisque $A * G$ est une extension normalisante libre de A et que A admet un anneau classique de fractions artinien, le corollaire 5.3 de [12] montre que $A * G$ admet un anneau classique de fractions artinien. Le théorème 3 de [11] montre alors que $A^G = e(A * G)e$ donc aussi A^G admet un anneau classique de fractions artinien. Il en résulte que les idéaux premiers minimaux de A^G sont tous de Goldie et forment un ensemble fini $\{p_1, p_2, \ldots, p_n\}$ et que l'on a $\mathscr{C}_{A^G}(0) = \mathscr{C}_{A^G}((\tilde{N})^G)$. Le radical premier $\tilde{N}$ de A est G-invariant et par suite G agit sur l'anneau semi-premier de Goldie $A/_{\tilde{N}}$. Comme $|G|^{-1} \in A$, on peut démontrer que l'on a : $(A/_{\tilde{N}})^G = A^G + \tilde{N}/_{\tilde{N}}$, et le théorème de Kharchenko implique l'inclusion :

$$\mathscr{C}_{(A/\widetilde{N})^G}(0) \subset \mathscr{C}_{A/\widetilde{N}}(0) .$$

Soit $c \in \mathscr{C}_{A^G}((\widetilde{N})^G)$. Comme $\varphi : A^G/_{(\widetilde{N})^G} \longrightarrow A^G + \widetilde{N}/_{\widetilde{N}} = (A/_{\widetilde{N}})^G : r + (\widetilde{N})^G \longmapsto r + \widetilde{N}$, est un isomorphisme d'anneaux, on a : $c + \widetilde{N} \in \mathscr{C}_{(A/_{\widetilde{N}})^G}(0) \subset \mathscr{C}_{A/_{\widetilde{N}}}(0)$. Par suite, il vient $c \in \mathscr{C}_A(\widetilde{N})$. On a donc démontré l'inclusion $\mathscr{C}_{A^G}(0) = \mathscr{C}_{A^G}((\widetilde{N})^G) \subset \mathscr{C}_A(0) = \mathscr{C}_A(\widetilde{N})$ et il en résulte que l'anneau classique des fractions $Q(A^G)$ de A^G est un sous-anneau de l'anneau classique des fractions $Q(A)$ de A. (Il est même possible de démontrer que $Q(A^G) = Q(A)^G$).

Le $(A/_I , A^G)$-bimodule $M = A/_I$ possède les propriétés suivantes :

a) M est sans torsion comme A^G-module à droite. En effet soient $m = r + I \in M$ et $c \in \mathscr{C}_{A^G}(0)$ tels que $mc = 0$; on obtient alors $rc \in I$ et comme d'après ce qui précède on a $\mathscr{C}_{A^G}(0) \subset \mathscr{C}_A(I)$, il vient $m = 0$.

De plus M est fidèle comme A^G-module à droite : $r_{A^G}(M) = I \cap A^G = \widetilde{I} \cap A^G = 0$.

b) A^G a un anneau classique de fractions artinien.

c) Puisque $A/_I$ a un anneau classique de fractions artinien, M considéré comme $A/_I$-module à gauche est de rang réduit fini et est sans torsion.

La proposition 1 montre alors que l'on a le principe d'additivité du rang et du rang réduit pour I tel qu'il est énoncé dans la partie 2) de la proposition 4.

EXEMPLES.

1. Soient, k un corps commutatif de caractéristique 0 et $A_n(k)$ la n-ième k-algèbre de Weyl. La k-algèbre $T = \left\{\begin{pmatrix} a & b \\ 0 & a \end{pmatrix} \middle| \, a,b \in A_n(k)\right\}$ est une k-algèbre noethérienne de type fini non semi-première, GK-homogène avec $\text{GK}\dim T = 2n$; les seuls idéaux (bilatères) de T sont : 0, $P = \begin{pmatrix} 0 & A_n(k) \\ 0 & 0 \end{pmatrix}$ et T. L'anneau classique des fractions de T est $S = \left\{\begin{pmatrix} s & t \\ 0 & s \end{pmatrix} \middle| \, s,t \in D_n(k)\right\}$ où $D_n(k)$ est le corps des fractions de $A_n(k)$. $A = T \times T$ est un k-algèbre de type fini, noethérienne, GK-homogène de GK-dimension 2n, non semi-première et non à I.P. A admet pour anneau classique des fractions la k-algèbre artinienne $Q(A) = S \times S$. L'idéal $I = 0$ de A est donc un idéal de Small de A GK-homogène dans A. On considère le k-automorphisme σ de A défini par : $\left(\begin{pmatrix} a & b \\ 0 & a \end{pmatrix}, \begin{pmatrix} a' & b' \\ 0 & a' \end{pmatrix}\right)^\sigma = \left(\begin{pmatrix} a & -b \\ 0 & a \end{pmatrix}, \begin{pmatrix} a' & b' \\ 0 & a' \end{pmatrix}\right)$. Le sous-groupe G de $\text{Aut}_k A$ engendré par σ est : $G = \{\text{id}_A , \sigma\}$. On a $|G|^{-1} \in A$ et en identifiant $A_n(k)$ à

$\left\{\begin{pmatrix} a & 0 \\ 0 & a \end{pmatrix}, a \in A_n(k)\right\}$ on obtient : $A^G = A_n(k) \times T$. A^G est une k-algèbre noethérienne de type fini, GK-homogène de GK-dimension 2n, non semi-première et non à I.P., qui admet pour anneau classique des fractions la k-algèbre artinienne $Q(A^G) = Q(A)^G = D_n(k) \times S$. $\mathrm{Spec}_{\min} A^G = \{p_1, p_2\}$ avec $p_1 = A_n(k) \times P$ et $p_2 = 0 \times T$.
On vérifie que l'on a conformément à la proposition 4, les relations :

$$\mathrm{rk}({}_A A) = 2 = \mu_1 \,\mathrm{rk}\, A^G/_{p_1} + \mu_2 \,\mathrm{rk}\, A^G/_{p_2} \quad \text{avec} \quad \mu_1 = \mu_2 = 1\,,$$
$$\mathrm{rrk}({}_A A) = 4 = \nu_1 \,\mathrm{rk}\, A^G/_{p_1} + \nu_2 \,\mathrm{rk}\, A^G/_{p_2} \quad \text{avec} \quad \nu_1 = \nu_2 = 2\,.$$

De plus on a ici la propriété d'équidimensionalité :

$$\mathrm{GK\,dim}\, A^G/_{p_1} = \mathrm{GK\,dim}\, A^G/_{p_2} = 2n\,.$$

2. Le localisé de $\mathbb{Z}$ par rapport à l'idéal $3\mathbb{Z}$ étant noté $\mathbb{Z}_{3\mathbb{Z}}$, on considère l'anneau noethérien à droite (mais non à gauche) $A = \begin{pmatrix} \mathbb{Z}_{3\mathbb{Z}} & \mathbb{Q} \\ 0 & \mathbb{Q} \end{pmatrix}$; A admet pour anneau classique de fractions (à droite et à gauche) $Q(A) = \begin{pmatrix} \mathbb{Q} & \mathbb{Q} \\ 0 & \mathbb{Q} \end{pmatrix}$; par suite $I = 0$ est un idéal de Small de A, G-invariant où $G = \{id_A, \sigma\}$ est le sous-groupe de Aut A engendré par l'automorphisme σ de A induit par la conjugaison par $\begin{pmatrix} 1 & 0 \\ 0 & -1 \end{pmatrix}$.
On a : $A^G = \begin{pmatrix} \mathbb{Z}_{3\mathbb{Z}} & 0 \\ 0 & \mathbb{Q} \end{pmatrix}$ et $\mathrm{Spec}_{\min} A^G = \{p_1, p_2\}$ avec $p_1 = \begin{pmatrix} \mathbb{Z}_{3\mathbb{Z}} & 0 \\ 0 & 0 \end{pmatrix}$ et $p_2 = \begin{pmatrix} 0 & 0 \\ 0 & \mathbb{Q} \end{pmatrix}$
Conformément à la proposition 4, on a les relations :

$$r_k({}_A A) = 2 = \mu_1 \,\mathrm{rk}\, A^G/_{p_1} + \mu_2 \,\mathrm{rk}\, A^G/_{p_2} \quad \text{avec} \quad \mu_1 = \mu_2 = 1\,.$$
$$\mathrm{rrk}({}_A A) = 3 = \nu_1 \,\mathrm{rk}\, A^G/_{p_1} + \nu_2 \,\mathrm{rk}\, A^G/_{p_2} \quad \text{avec} \quad \nu_1 = 2 \quad \text{et} \quad \nu_2 = 1\,.$$

Ici on n'a pas la propriété d'équidimensionalité relative à la K dim car :

$$\mathrm{K\,dim}\, A^G/_{p_1} = 0 \neq \mathrm{K\,dim}\, A^G/_{p_2} = 1\,.$$

Cela vient de ce que A n'est pas Krull-homogène.

4. EQUIDIMENSIONALITE.

Proposition 5. — *Soient,* A *une* k-*algèbre noethérienne et* G *un groupe fini de* k-*automorphismes de* A *tel que* $|G|^{-1} \in A$. *Alors pour tout idéal* I *GK-homogène dans* A, *on a :*

$$\mathrm{GK}\dim A^G/_p = \mathrm{GK}\dim A^G/_{I\cap A^G}$$

pour tout idéal premier p *de* A^G *minimal dans l'ensemble des idéaux premiers de* A^G *contenant* $I \cap A^G$.

Démonstration. — La démonstration est basée sur le lemme suivant que nous avons donné dans [4].

Lemme 6. — *Soient,* A *et* B *deux* k*-algèbres noethériennes et* ${}_BM_A$ *un* (B,A)*-bimodule de type fini à droite et à gauche. Alors les propriétés suivantes sont équivalentes :*

(i) ${}_BM$ *est* GK*-homogène et* $\mathrm{GK}\dim({}_BM) = \mu$;

(ii) M_A *est* GK*-homogène et* $\mathrm{GK}\dim(M_A) = \mu$;

(iii) *Pour tout* $Q \in \mathrm{Ass}({}_BM)$, *on a* $\mathrm{GK}\dim B/_Q = \mathrm{GK}\dim({}_BM) = \mu$;

(iv) *Pour tout* $P \in \mathrm{Ass}(M_A)$, *on a* $\mathrm{GK}\dim(A/_P) = \mathrm{GK}\dim(M_A) = \mu$.

Si ces conditions sont réalisées alors $A/_{r_A(M)}$ *et* $B/_{\ell_B(M)}$ *sont des* k*-algèbres* GK*-homogènes de* GK*-dimension* μ.

Démontrons la proposition 5. Il existe un monomorphisme de (A,A)-bimodules de $A/_{\tilde{I}}$ dans $\bigoplus_{x\in X} A/_{I^x}$ où X est un système de représentants dans G des classes à droite modulo le stabilisateur de I. Pour tout $x \in X$, $A/_{I^x}$ est une k-algèbre noethérienne GK-homogène isomorphe à $A/_I$: $\mathrm{GK}\dim A/_{I^x} = \mathrm{GK}\dim A/_I = \mu$. En tant que $A/_{I^x}$-module à gauche, donc aussi en tant que A-module à gauche, $A/_{I^x}$ est GK-homogène de GK-dimension μ pour tout $x \in X$. Il en résulte qu'en tant que A-module à gauche $A/_{\tilde{I}}$ est GK-homogène de GK-dimension μ. Comme G opère sur la k-algèbre noethérienne $A/_{\tilde{I}}$ et comme $|G|^{-1} \in A/_{\tilde{I}}$, la k-algèbre des points fixes $(A/_{\tilde{I}})^G$ est noethérienne et $A/_{\tilde{I}}$ est un $(A/_{\tilde{I}})^G$-module à droite fidèle de type fini. En appliquant le lemme 6 précédent au (A , $(A/_{\tilde{I}})^G$)-bimodule $A/_{\tilde{I}}$, on en déduit que $(A/_{\tilde{I}})^G$ est une k-algèbreGK-homogène de GK-dimension μ. Comme $|G|^{-1} \in A$ on a :

$$(A/_{\tilde{I}})^G \simeq A^G/_{\tilde{I}\cap A^G} = A^G/_{I\cap A^G} .$$

L'équidimensionalité résulte alors d'un résultat de W. Borho [2].

Remarques.

1. Si A est une k-algèbre noethérienne de type fini et de GK-dimension finie, la proposition 5 résulte du corollaire 6 que nous avons donnée dans [4].

2. Dans [9] S. Montgomery définit une équivalence $\sim$ dans Spec A^G. Si p et q sont deux idéaux premiers équivalents de A^G, il existe $P \in$ Spec A tel que p et q soient minimaux au-dessus de $P \cap A^G$ dans Spec A^G. Comme P est GK-homogène dans A on obtient en appliquant la proposition 5, le corollaire suivant.

Corollaire 7. — *Soient* A *une* k*-algèbre noethérienne et* G *un groupe fini de* k*-automorphismes de* A *tel que* $|G|^{-1} \in A$. *Si dans* Spec A^G, *on a :* $p \sim q$ *alors* GK dim $A^G/_p$ = GK dim $A^G/_q$.

Si A est à I.P., ce corollaire résulte du théorème 2 de [10]. Dans [10], il existe un exemple d'application de ce corollaire qui échappe au domaine d'application du théorème 2 de [10].

Proposition 8. — *Soient,* A *un anneau noethérien à* I.P. *et* G *un groupe fini d'automorphismes de* A *tel que* $|G|^{-1} \in A$. *Si* I *est un idéal Krull-homogène dans* A, *alors on a :* K dim $A^G/_p$ = K dim $A^G/_{I \cap A^G}$ *pour tout idéal premier* p *de* A^G *minimal dans l'ensemble des idéaux premiers de* A^G *contenant* $I \cap A^G$. *En particulier si dans* Spec A^G *on a* $p \sim q$ *alors* K dim $A^G/_p$ = K dim $A^G/_q$.

Sous les hypothèses de cette proposition 8, il est facile de vérifier que la K dim vérifie l'analogue du lemme 6 et du résultat de Borho [2] utilisé dans le démonstration de la proposition 5. La démonstration de la proposition 8 suit alors les mêmes lignes que celles de la proposition 5.

Remarque. — Si A est une k-algèbre à I.P. noethérienne et de type fini, la dernière partie de la proposition redonne un résultat de J. Alev, car dans ce cas pour tout idéal premier p de A^G, la dimension de Krull habituelle dim $A^G/_p$, définie à partir des chaînes d'idéaux premiers, coïncide avec K dim $A^G/_p$.

5. APPLICATION DU PRINCIPE D'ADDITIVITE AUX EXTENSIONS TRIANGULAIRES.

Suivant [6], un anneau B est une extension triangulaire d'un anneau A, si A est un sous-anneau de B (A et B admettant le même élément unitaire 1), et s'il existe une suite b_1 ($b_1 = 1$), $b_2, \ldots, b_n$ d'éléments de B telle que pour tout $i \in \{1,2,\ldots,n\}$ on ait $\sum_{j=1}^{i} Ab_j = \sum_{j=1}^{i} b_j A$, avec $\sum_{j=1}^{n} Ab_j = \sum_{j=1}^{n} b_j A = B$. L'intérêt de cette généralisation de la notion d'extension normalisante est montré dans [6]. Un des obstacles pour étendre aux extensions triangulaires les résultats connus des extensions normalisantes est le suivant : B étant une extension triangulaire de A et P un idéal premier de B, $P \cap A$ n'est pas nécessairement un idéal semi-premier de A. Néanmoins le principe d'additivité rappelé précédemment permet de généraliser le théorème 3 de R.B. Warfield [14] obtenu pour les extensions normalisantes, comme suit :

Proposition 9. — *Soient* B *une extension triangulaire de* A *et* P *un idéal premier de* B *de Goldie à gauche. S'il existe au moins un idéal premier* p *de* A *minimal au-dessus de* $P \cap A$ *qui soit de Goldie à droite, les propriétés suivantes sont vérifiées :*

1) *Les idéaux premiers de* A *minimaux au-dessus de* $P \cap A$, *forment un ensemble fini* $\{p_1, p_2, \ldots, p_m\}$ *et tous les* $A/_{p_i}$ *sont isomorphes.*

2) *Pour tout* $i \in \{1,2,\ldots,m\}$, $\text{rk } A/_{p_i}$ *divise* $\text{rk } B/_P$.

Remarque. — La partie 1) de la proposition n'est citée que pour mémoire ; en effet, il s'agit du "Cutting down" de [6].

Démonstration. — On peut supposer que $P = 0$. L'anneau classique des fractions à gauche de B est noté C ; C est un anneau simple artinien. Soit ${}_C X_A$ un sous-(C,A)-bimodule propre maximal du (C,A)-bimodule C. On pose conformément aux notations de [6] : ${}^iX = \bigcap_{j=1}^{i} X^{\bullet}.b_j$ avec $X^{\bullet}.b_j = \{q \in C \mid qb_j \in X\}$. ${}^iX$ est un sous-(C,A)-bimodule de C.

Soit $x \in {}^nX$. On a alors $xb_j \in X$ pour $j = 1,\ldots,n$ ce qui implique $x \sum_{j=1}^{n} b_j A \subset X$ donc $xB \subset X$. Il en résulte que $BxB \subset X$. $B \cap BxB$ est un idéal bilatère de l'anneau premier de Goldie à gauche B. Comme $C(X \cap B) \subset CX \subset X \subsetneqq C$, $X \cap B$ est un idéal à gauche de B non essentiel, ce qui implique que $B \cap BxB$ n'est pas essentiel à gauche dans B, donc que $B \cap BxB = 0$ et par suite $BxB = 0$ car B est un sous-B-module à gauche essentiel de ${}_BC$. Par conséquent $x = 0$, ce qui prouve ${}^nX = 0$. On a donc la chaîne suivante de (C,A)-sous-bimodules de C :

$$0 = {}^nX \subset {}^{n-1}X \subset \ldots \subset {}^1X = X \subsetneqq C = {}^0X\ .$$

Pour $i \in \{1,2,\ldots,n\}$, on considère $\theta_i = {}^{i-1}X/{}^iX \to C/_X$: $[q + {}^iX] \longmapsto [qb_i + X]$. On vérifie que θ_i est un homomorphisme de groupe additif qui est injectif. On vérifie que $Z \longmapsto \theta_i(Z)$ est une injection strictement croissante du treillis des (C,A)-sous-bimodule de ${}^{i-1}X/{}^iX$ dans le treillis des (C,A)-sous-bimodules de $C/_X$. Donc, soit ${}^{i-1}X/{}^iX = 0$, soit ${}^{i-1}X/{}^iX$ est un (C,A)-bimodule simple ; dans ce dernier cas on pose $p_i = r_A({}^{i-1}X/{}^iX)$ et on vérifie que p_i est un idéal premier de A. Soit $I = \{i \in \{1,2,\ldots,n\} \mid {}^iX \subsetneqq {}^{i-1}X\}$; il est clair que tout idéal premier minimal de A est l'un des p_i pour $i \in I$. Si on pose pour $i \in I$, $M_i = {}^iX \in B$, les sous(B,A)-bimodules M_i de B permettent de construire une chaîne du même type que celle considérée dans la démonstration de la proposition 2 (voir [4]) , avec $p_i = r_A(M_{i-1}/M_i)$.

Pour $i \in I$, on a :

$$p_i = \{a \in A \mid {}^{i-1}Xa \subset {}^iX\} = \{a \in A \mid {}^{i-1}Xab_i \subset X\}\ ;$$

On considère alors $\psi_i : A \longrightarrow A/_{p_i}$: $a \longmapsto [a_i + p_i]$ avec $a_i \in A$ tel que

$b_i a - a_i b_i \in \sum_{j=1}^{i-1} Ab_j = \sum_{j=1}^{i-1} b_j A$; on vérifie que ψ_i est un homomorphisme surjectif

d'anneaux. On a ${}^{i-1}Xb_i \not\subset X$ (pour $i \in I$) et par suite le (C,A)-bimodule ${}^{i-1}Xb_i + {}^{i-1}Xb_{i-1} + \ldots + {}^{i-1}Xb_1$ n'est pas inclus dans X. Soit $a \in \text{Ker } \psi_i$. Alors

$\psi_i(a) = [a_i + p_i] = [0 + p_i]$ implique ${}^{i-1}Xa_i b_i \subset X$ et ${}^{i-1}Xb_i a \subset X$. On obtient alors $({}^{i-1}Xb_i + {}^{i-1}Xb_{i-1} + \ldots + {}^{i-1}Xb_1)a \subset X$ et par suite puisque le (C,A)-bimodule $C/_X$ est simple il en résulte $a \in p_1 = r_A(C/_X)$ et finalement $\operatorname{Ker} \psi_i \subset p_1$ (pour $i \in I$). Il est alors facile de vérifier que $\operatorname{Ker} \psi_i = p_1$ donc que pour $i \in I$ on a $A/_{p_1} \simeq A/_{p_i}$.
La remarque suivant la proposition 2 implique : $\operatorname{rk} B = (\sum_{i \in I} \nu_i) \operatorname{rk} A/_{p_1}$.

Remarque. — Dans une correspondance privée, B. Lemonnier m'a indiqué que dans la démonstration précédente, on peut encore considérer la chaîne suivante de sous-(C,A)-bimodules de C :

$$0 \subsetneq U^1 = U \subset U^2 \subset \ldots \subset U^{n-1} \subset U^n = C$$

où U est un sous-(C,A)-bimodule non nul minimal de C et où pour $i \in \{1,2,\ldots,n\}$ on a $U^i = \sum_{j=1}^{i} U b_j$.

5. BIBLIOGRAPHIE.

[1] J.A. BEACHY, Rings with finite reduced rank, Comm. Algebra, 10 (14), (1982), 1517-1536.

[2] W. BORHO, "On the Joseph - Small additivity principle for Goldie ranks", Compositio Mathematica 47 (1), (1982), 3-29.

[3] W. BORHO and H. KRAFT, "Über die Gelfand-Krillov Dimension", Math. Annalen 220, (1976), 1-24.

[4] A. HUDRY, "Sur le principe d'additivité du rang pour des bimodules, II", Proceedings Colloque d'algèbre non commutative, Luminy (1984).

[5] A. JOSEPH et L.W. SMALL, "An additivity principle for Goldie rank", Israël J. of Math. 31 (2), (1978), 105-114.

[6] B. LEMONNIER, Thèse, Université de Poitiers, 387, (1984).

[7] T.H. LENAGAN, "Gelfand-Kirillov Dimension and affine P.I. rings", Comm. Algebra, 10 (1), (1982), 87-92.

[8] M. LORENZ, S. MONTGOMERY, L.W. SMALL, "Prime ideals in fixed rings, II", Comm. in Algebra, 10 (5), (1982), 449-455.

[9] S. MONTGOMERY, "Prime ideals in fixed rings", Comm. in Algebra, 9 (4), (1981), 423-449.

[10] S. MONTGOMERY and L.W. SMALL, "Integrality and prime ideals in fixed rings of P.I. rings", J. of pure and applied Algebra 31, (1984), 185-190.

[11] L.W. SMALL, "Orders in artinian rings, II", J. of Algebra, 9, (1968), 266-273.

[12] J.T. STAFFORD, "Noetherian full quotient rings", Proc. London, Math. Soc. (3), 44 (1982), 385-404.

[13] J.T. STAFFORD, "The Goldie rank of a module, preprint Univ. of Leeds, (1983), 1-35.

[14] R.B. WARFIELD, "Prime ideals in rings extensions", J. London Math. Soc. 28 (1983), 453-460.

Cohomology of $\mathbb{P}^m_{(2)}$.

Lieven Le Bruyn (*)

0. Some motivation.

A basic theme in noncommutative ring theory is the attempt to generalize results of commutative ring theory to noncommutative rings. A good example of this theme is algebraic geometry for p.i. rings, which currently is quite popular.
This note can be viewed as an attempt to generalize Serre's result on the cohomology of projective space :

Theorem 0. (Serre, Cohomology of $\mathbb{P}^m$)

(1): $F[X_0, \ldots, X_m] \simeq \oplus H^0(\mathbb{P}^m, \underline{\vartheta}_{\mathbb{P}^m}(n))$

(2): $H^i(\mathbb{P}^m, \underline{\vartheta}_{\mathbb{P}^m}(n)) = 0$ for all n and for $0 < i < m$.

(3) : The dimension of the F-vectorspace

$$H^m(\mathbb{P}^m, \underline{\vartheta}_{\mathbb{P}^m}(-m-1-n))$$

is equal to $\binom{m+n}{n}$

to quaternionic projective space, or rather spaces. Of course, we have to define what we mean by this.

Let us start by considering the identities of 2 by 2 matrices, i.e.

$$\Lambda \simeq \mathbb{G}_{m,2}/I$$

for some twosided ideal I of $\mathbb{G}_{m,2}$. Here, $\mathbb{G}_{m,2}$ is the ring of m generic 2 by 2 matrices. It is the F-sub-algebra of

$$M_2(\mathcal{P}_m) = M_2(F[X_{11}(l), X_{12}(l), X_{21}(l), X_{22}(l) : 1 \leq l \leq m])$$

generated by the so called generic matrices

$$X_l = \begin{pmatrix} X_{11}(l) & X_{12}(l) \\ X_{21}(l) & X_{22}(l) \end{pmatrix}$$

We are interested in 2-dimensional representations of Λ, i.e. F-algebra morphisms

$$\varphi : \Lambda \to M_2(F)$$

This study can be seen as the description of the set of solutions in the matrix-variables $X_1, \ldots, X_m$ to the ideal of relations I, i.e. the topic of interest of what might be called (some day, hopefully) noncommutative algebraic geometry.
Clearly, we are not interested in all representations but more in a description for the equivalence classes of the relation

$$\varphi \tilde{\varphi}' \quad \text{iff} \quad \exists \alpha \in \mathrm{Aut}_F(M_2(F)) : \varphi = \alpha' 0 \varphi$$

The main difficulty is that there is no scheme parametrizing these equivalence classes, cfr. [Ar]. This obstruction motivated Artin and Procesi to reformulate the object of study : characterize all equivalence classes of 2-dimensional semi-simple representations (and their irreducible components) of the affine Λ.
This can be done in the following way, see the work of Procesi [P1] and Artin-Schelter [A3].
Consider our algebra $\Lambda = \mathbb{G}_{m,2}/I$ then it is easily seen that

$$M_2(P_m)/M_2(P_m).I.M_2(P_m) \simeq M_2(S)$$

where $S = P_m/J$ where J is the ideal generated by the entries of the matrices in I.
We have the situation :

$$\begin{array}{ccc} \mathbb{C}_{m,2} & \longrightarrow & M_2(P_m) \\ \downarrow & & \downarrow \\ \Lambda & \xrightarrow{\pi} & M_2(S) \end{array}$$

With $T(\Lambda)$ we will denote the sub F-algebra of $M_2(S)$ generated by the image of Λ and $c(\Lambda)$ where $c(\Lambda)$ id the sub F-algebra of S generated by all coefficiënts of characteristic polynomials of elements of $\pi(\Lambda)$.
It is easy to see that there are F-algebras epimorphisms

$$\begin{array}{ccc} \mathbb{T}_{m,2} & \longrightarrow & T(\Lambda) \\ \downarrow & & \downarrow \\ R_{m,2} & \longrightarrow & c(\Lambda) \end{array}$$

where $\mathbb{T}_{m,2}$ is $T(\mathbb{G}_{m,2})$ the so called trace ring of m generic 2 by 2 matrices.
For any affine F-algebra Γ we will denote by $\mathrm{Max}(\Gamma)$ (resp. $\mathrm{Spec}(\Gamma)$) the set of all twosided maximal (resp. prime) ideals of Γ. The main result can now be stated as follows

Theorem. (Artin - Schelter, Th. 3.20.)

(1) There is a one to one correspondence between $\mathrm{Max}(c(\Lambda))$ and equivalence classes of 2-dimensional semi-simple representations Λ.

(2) $\mathrm{Max}(T(\Lambda))$ consists of couples (φ, φ_i) when $\varphi : \Lambda \to M_2(F)$ is a representant of a 2-dimensional semi-simple representation of Λ and φ_i is an irreducible factor of φ.

Therefore, the study of solutions in m 2 by 2 matrix variables to an ideal of relations I of $\mathbb{G}_{m,2}$ amounts to the study of the maximal (or prime) ideal spectrum of $T(\mathbb{G}_{m,2}/I)$. The epimorphism

$$\mathbb{T}_{m,2} \to T(\mathbb{G}_{M,2}/I)$$

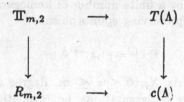

induces continuous maps :

$$\mathrm{Spec}T(\Lambda) \to \mathrm{Spec}\mathbb{T}_{m,2}$$

$$\mathrm{Max}T(\Lambda) \to \mathrm{Max}\mathbb{T}_{m,2}$$

So one can view $\mathrm{Spec}\mathbb{T}_{m,2}$ as a quaternionic generalization of affine m-space. But, from the work of Procesi [P2] it follows that

$$\mathbb{T}_{m,2} = \mathbb{T}^0_m[Tr(X_1), \ldots, Tr(X_m)]$$

where $\mathbb{T}^0_m$ is the subalgebra generated by the generic trace zero matrices

$$X_i^0 = X_i - \frac{1}{2}Tr(X_i)$$

So,

$$\mathrm{Spec}\mathbb{T}_{m,2} = \mathrm{Spec}\mathbb{T}^0_m \times A^m$$

and so, the real noncommutative (and hard) part of the problem is the description of Spec or Max of $\mathbb{T}^0_m$, which we will call quaternionic m-space and denote by $A^m_{(2)}$. Clearly, one can also study the projective version of these questions. In that case one starts off with a positively graded F-algebra Λ satisfying the identities of 2 by 2 matrices. which is generated by a finite number of homogeneous elements of degree one, i.e. we have a gradation preserving epimorphism

$$\varphi : \mathbb{G}_{m+1,2} \to \Lambda$$

if we give every generic matrix X_i, $0 \leq i \leq m$, degree one. Then we want to study 2-dimensional projective representations, i.e. gradation preserving F-algebra morphisms

$$\varphi : \Lambda \to \Delta$$

where Δ is a graded central simple algebra of dimension 4 over its center. Since F is algebraically closed, it follows from [NV] that

$$\Delta \simeq M_2\left(F[y, y^{-1}]\right)(\sigma_1, \sigma_2)$$

where $\deg(y) = 1$ or 2; $\sigma_1, \sigma_2 \in \mathbb{N}$ and $\mathbb{Z}$-gradation on Δ is defined by

$$\Delta_i = \begin{bmatrix} F[y, y^{-1}]_i & F[y, y^{-1}]_i \\ F[y, y^{-1}]_{i+\sigma_2-\sigma_1} & F[y, y^{-1}]_i \end{bmatrix}$$

We want to study equivalence classes, for the relation that $\varphi\tilde{\varphi}'$ is a gradation preserving automorphism between the target rings α s.t. $\varphi' = \alpha 0 \varphi$, of semisimple representations, i.e. such that $\varphi(\Lambda)$ is a graded semi-simple ring [NV].

Combining the argument of the affine case given above with ideas of [LVV], it is possible to show that this study amounts to a description of

$$\mathrm{Proj}\,(T(\Lambda))$$

where $T(\Lambda)$ is of course positively graded and Proj consists of all graded prime ideals of $T(\Lambda)$ not containing $T(\Lambda)_+ = \bigoplus_{i\geq 1} T(\Lambda)_i$.

Again, $\mathrm{Proj}(\mathbb{T}_{m+1,2})$ can be considered as a quaternionic generalization of projective space but since $\mathbb{T}_{m+1,2} = \mathbb{T}^0_{m+1}[Tr(X_0),\ldots,Tr(X_m)]$ we christen the hard, noncommutative part of this problem, i.e. $\mathrm{Proj}(\mathbb{T}^0_{m+1})$, quaternionic m- space and we denote it by $\mathbb{P}^m_{(2)}$.

An important difference with the commutative case is that the rings $\mathbb{T}^0_{m+1}$ are allmost never regular, [L1]. Moreover, M. Van den Bergh has shown that for $m > 3$, $\mathbb{T}^0_{m+2}$ cannot be obtained as an epimorphic image of a positively graded F-algebra of finite global dimension satisfying the identities of 2 by 2 matrices.
If we drop the assumption on having the same p.i.-degree, such an F- algebra does exist [L2].
For, take the iterated Õre extension

$$Cl_{m+1} = F[a_{ij} : 0 \leq i < j \leq m][a_0][a_1,\sigma_1,\delta_1]\ldots[a_m,\sigma_m,\delta_m]$$

where for each $i < j, \delta_j(a_i) = 2a_{ij}$ and $\sigma_j(a_i) = -a_1$ and trivial action on the other variables.
Then, sending a_i to X^0_i and a_{ij} to $\frac{1}{2}Tr(X^0_i X^0_j)$ we obtain an epimorphism

$$\varphi_{m+1} : Cl_{m+1} \to \mathbb{T}^0_{m+1}$$

Therefore, we can view $\mathrm{Proj}(Cl_{m+1})$ as a sort of regular quaternionic m-space, $\mathbb{P}^m_{reg}$.
A pleasant property is that, whereas prime ideals of Cl_{m+1} can split up wildly over primes of the central subring

$$S_{m+1} = F[a_{ij} : 0 \leq i \leq j \leq m]$$

where $a_{ii} = (a_i)^2$, graded prime ideals lie uniquely i.e. $\mathbb{P}^m_{reg}$ is homeomorphic, as topological spaces to projective $\binom{m+1}{2} - 1$-space.
In this note we aim to construct structure sheafs and to compute the cohomology of both spaces. The main result will be

Theorem 8. (cohomology of $\mathbb{P}^m_{(2)}$)

(1) : $\mathbb{T}^0_{m+1} \simeq \bigoplus_{n\in\mathbb{Z}} H^0(\mathbb{P}^m_{(2)}, \underline{\vartheta}_{\mathbb{P}^m_{(2)}}(n))$

(2) : $H^i(\mathbb{P}^m_{(2)}, \underline{\vartheta}_{\mathbb{P}^m_{(2)}}(n)) = 0$ for all n and all $0 < i < 3m-1$

(3) : The dimension of the F-vector space

$$H^{3m-1}\left(\mathbb{P}^m_{(2)}, \underline{\vartheta}_{\mathbb{P}^m_{(2)}}(-3m-3-n)\right)$$

is equal to the number of standard Young tableaux of length ≤ 3 having n boxes. The proof of it relies on three facts :

(1) : We can cover $\mathbb{P}^m_{(2)}$ and $\mathbb{P}^m_{\text{reg}}$ with affine open sets such that the sections of the structure sheaves contain units of degree one.

(2) : the ring of generic trace zero 2 by 2 matrices, $\mathbb{T}^0_{m+1}$, is a Cohen-Macauley module, cfr. [L3].

(3) : The Poincaré series of $\mathbb{T}^0_{m+1}$ satisfies a functional equation, [L1].

We believe that from this point on, it should be possible to develop a "quaternionic geometry". Modesty forces us to leave this topic to people with geometrical expertise. Hopefully, these results will be of some value to them.

Acknowledgement.

This paper is a continuation of my talk forin the séminaire P. Dubreuil, M.P. Malliavin in november '84 on Cohen-Macauleyness of $\mathbb{T}_{m,2}$. I like to thank M.P. Malliavin for her kind invitation.

This paper was written while nursing our baby, which I hereby thank for her patience.

1. The scheme structure.

In this section we will calculate the local structure of the schemes $\mathbb{P}^m_{(2)}$ and $\mathbb{P}^m_{\text{reg}}$. We will start off with the last one. First, we recall some properties of the so called generic Clifford algebras. For proofs and more details, the reader is referred to [L3] or [L4].

(a) : Generic Clifford Algebras.

Let R be any commutative f-algebra. An m-ary quadratic form over the ring R is a polynomial f in m variables over R which is homogeneous of degree two, i.e.

$$f = \Sigma \alpha_{ij} X_i X_j \in R[X_1, \ldots, X_m]$$

with $\alpha_{ij} = \alpha_{ij} \in R$. So, f determines uniquely a symmetric m by m matrix with coefficients in R

$$M_f = (\alpha_{ij})_{ij} \in M_m(R)$$

Let R^m denote the standard free R-module of rank m with basis $e_1, \ldots, e_m$ the unit vectors. The quadratic form f gives rise to a quadratic map

$$Q_f : R^m \to R$$

defined by sending a column m-tuple $x = (x_1, \ldots, x_m)^\tau$ to $Q_f(x) = x^\tau . M_f . x$. To any m-ary quadratic form f over R one can associate its Clifford algebra $Cl(R, f)$ which is defined to be the quotient of the tensor algebra of the R-module R^m modulo the twosided ideal generated by all elements

$$x \otimes x - Q_f(x)$$

where $x \in R^m$. If we give the tensor algebra the usual $\mathbb{Z}$- gradation, then $x \otimes x$ is homogeneous of degree 2 whereas $_f(x)$ is of degree zero. This entails that the Clifford algebra $Cl(R, f)$ has an induced $\mathbb{Z}/2\mathbb{Z}$- gradation, i.e.

$$Cl(R, f) = C_0 \oplus C_1$$

with $C_i . C_j \subset C_k$ where $k \equiv i + j \bmod 2$.
We will now introduce a noncommutative F-algebra Cl_m which is generic in the sense that every Clifford algebra of an m-ary quadratic form over F can be obtained as a specialization of Cl_m.
Let S_m be the homogeneous coordinate ring of the variety of symmetric m by m matrices with entries in F, i.e. S_m is the commutative polynomial ring

$$F[a_{ij} : 1 \leq i \leq j \leq m]$$

in $\binom{m+1}{2}$ indeterminates. By f_m we will denote the following m-ary regular quadratic form over S_m

$$f_m(X_1,\ldots,X_m) = \sum_{i,j=1}^{m} a_{ij}X_iX_j$$

The mth generic Clifford algebra over F, Cl_m, is defined to be the Clifford algebra $Cl(S_m, f_m)$.
If $f = \Sigma\alpha_{ij}X_iX_j$ is any m-ary quadratic form over F, then specializinga_{ij} to α_{ij} gives an F- algebra epimorphism

$$\pi_f : Cl_m \to Cl(F,f)$$

It is possible to give a more concrete description of Cl_m. Consider the iterated Öre-extension

$$\Lambda_m = F[a_{ij} : 1 \leq i < j \leq m][a_1][a_2,\sigma_m,\delta_m]$$

where one defines for each $i < j$ that $\sigma_j(a_i) = -a_i$ and $\delta_j(a_i) = 2a_{ij}$ and trivial actions of σ_j and δ_j on the other indeterminates.
Using the universal property of Clifford algebras it is easy to check that $\Lambda_m \simeq Cl_m$.
Moreover, giving each of the variables a_{ij} degree 2 and the a_i degree one, one checks that Cl_m is a positively graded F- algebra generated by the m elements of degree one : $a_1,\ldots,a_m$.
Further, Cl_m has finite global dimension equal to $\binom{m+1}{2}$ and the p.i.-degree of the generic Clifford algebra Cl_m is equal to 2^α is the largest natural number $\leq \frac{m}{2}$.
Also, Cl_m is a maximal order and its center Z_m is equal to

$$\begin{array}{ll} S_m & \text{if } m \text{ is even} \\ S_m \oplus S_m.d & \text{if } m \text{ is odd} \end{array}$$

where

$$d = S_m(a_1,\ldots,a_m) = \sum_{\sigma\in S_m} \mathrm{sgn}(\sigma)a_{\sigma(1)}\ldots a_{\sigma(m)}$$

where s_m is the permutation group on m elements. Of course, one has $a_i^2 = a_{ii}$.
We will need a concrete description of Spec(Cl_m), i.e. the set of all twosided prime ideals of Λ equipped with the usual Zariski topology. Clearly, intersecting with S_m yields a continuous map

$$\varphi : \mathrm{Spec}(Cl_m) \to \mathrm{Spec}(S_m)$$

and since Cl_m is a finite module over S_m (even free of rank 2^m). This map is surjective.
The prime ideal spectrum of a commutative polynomial ring (such as S_m) may be assumed to be relatively well known. Therefore, describing Spec(Cl_m) essentially amounts to describing the fibers of φ.

Proposition 1. [13] If p is any prime ideal of S_m, the fiber $\varphi^{-1}(p)$ contains at most two elements.

Moreover, it is easy to see whether $\varphi^{-1}(p)$ has one or two elements. For let

$$\Pi(\mathcal{A}) = (\Pi(a_{ij}))_{i,j}$$

be the symmetric m by m matrix over the domain $S = S_m/p$. If the rank of $\Pi(\mathcal{A})$ is even, then $\varphi^{-1}(p)$ has just one element. If the rank of $\pi(\mathcal{A})$ is odd, $\pi(\mathcal{A})$ is congruent over the field of fractions of S, K to a matrix of the form

$$\begin{pmatrix} A & 0 \\ 0 & 0 \end{pmatrix}$$

where A is a symmetric invertible k by k matrix over K, $k = \text{rank}(\Pi(\mathcal{A}))$. Let

$$\delta = (-1)^{\binom{k}{2}}.\det(A)$$

then $\varphi^{-1}(n)$ has one element if $\delta \notin (K^*)^2$ and has two elements if $\delta \in (K^*)$?
In our situation where F is an algebraically closed field, the set of maximal ideals of S_m corresponds bijectively to the set of symmetric m by m matrices over F. In this case, the number of maximal ideals of Cl_m lying over a maximal ideal corresponding to a matrix $(\alpha_{ij})_{i,j}$ is equal to $1 + (\text{rank}(\alpha_{ij})\text{mod}2)$.

There is a main automorphism on Cl_m defined by sending a_i to $-a_i$ or in terms of $\mathbb{Z}/2\mathbb{Z}$-graded algebras by sending an element

$$x = x_0 \oplus x_1$$

$x \in Cl_m, x_i \in C_i$, to

$$x_0 \oplus (-x_1)$$

It is easy to verify, using the classical structure results of Clifford algebras over a field, see e.g. [Lam], that this automorphism fixes every prime ideal of Cl_m lying uniquely over S_m and permutes the two primes in the other case.

(b) : Local structure of $\mathbb{P}^m_{\text{reg}}$.

The generic Clifford algebras being positively graded F- algebras one can define

$$\mathbb{P}^m_{\text{reg}} = \text{Proj}(Cl_{m+1})$$

where Cl_{m+1} is the generic Clifford algebra generated by $a_0, \ldots, a_m$ and Proj (Cl_{m+1}) is the set of all twosided graded prime ideals of Cl_{m+1} not containing the positive part $(Cl_{m+1})_+ = \bigoplus_{i \geq 1} (Cl_{m+1})_i$ equipped with the induced Zariski topology.

Our first aim will be to prove that $\mathbb{P}^m_{\text{reg}}$ has the structure of a scheme, i.e. it can be covered by open sets which are homeomorphic to the prime ideal spectrum of certain F-algebras. Later we will define structure sheaves on $\mathbb{P}^m_{\text{reg}}$. For any central homogeneous element f of Cl_{m+1}, we denote

$$X_+(f) = \{P \in \mathbb{P}^m_{\text{reg}} : f \notin P\}$$

The open sets we choose are

$$\begin{array}{rll} I: & X_+(a_{ii}) & 0 \leq i \leq m \\ II: & X_+(2a_{ij} + a_{ii} + a_{jj}) & 0 \leq i < j \leq m \end{array}$$

We will treat every case separately.

Case I : By C_i we denote the graded localization of Cl_{m+1} at the central multiplicative set of homogeneous elements

$$\{a_{ii}^n : n \in \mathbb{N}\}$$

Then it is clear that C_i contains an homogeneous unit of degree one : a_i. Therefore :

$$C_i \simeq (C_i)_0[a_i, a_i^{-1}, \sigma]$$

where σ is the F-automorphism of $(C_i)_0$ determined by conjugation with a_i in C_i, i.e. for every $x \in (C_i)_0$:

$$\sigma(x) = a_i.x.a_i^{-1}$$

We will now calculate the part of degree zero.

Lemma 2: $(C_i)_0 \simeq Cl_m[x_1, \ldots, x_m]$

Proof.
Clearly, $(C_i)_0$ is generated by the elements

$$a_{kl}.a_{ii}^{-1}; a_k.a_i^{-1}$$

for all $0 \leq k <\leq m$ different from i. Let us compute the relations between them.

$$\begin{cases} a_k.a_i^{-1}.a_la_i^{-1} = -a_ka_la_{ii}^{-1} + a_{li}a_k^{-1}.a_i^{-1} \\ a_l.a_i^{-1}.a_k.a_i^{-1} = -a_la_ka_{ii}^{-1} + a_{ki}.a_{ii}^{-1}.a_l.a_i^{-1} \end{cases}$$

Therefore, if we denote

$$\begin{cases} A_k = a_k.a_i^{-1} & \text{for all } 0 \leq h \leq m \text{ different from } i \\ A_{kl} = \frac{1}{2}a_{kl}.a_{ii}^{-1} & \text{for all } 0 \leq h < l \leq m \text{ different from } i \\ A_{ki} = a_{ki}.a_{ii}^{-1} & \text{for all } 0 \leq h \leq m \text{ different from } i \end{cases}$$

we get the following relation for all $k < l$

$$(*) : A_k.A_l + A_l.A_k = -A_{kl} + A_{ki}A_l + A_{li}.A_k$$

Now, change the variables in the following way :

$$\begin{cases} B_k = \frac{1}{2}A_{ki} - A_k & \text{for all } 0 \leq h \leq m \text{ different from } i \\ B_{kl} = \frac{1}{2}A_{ki}A_{li} - A_{kl} & \text{for all } 0 \leq h <\leq m \text{ different from } i \end{cases}$$

then we obtain that

$$\begin{cases} B_kB_l = \dfrac{1}{4}A_{ki}A_{li} - \dfrac{1}{2}A_{ki}A_l - \dfrac{1}{2}A_{li}A_k + A_kA_l \\ B_lB_k = \dfrac{1}{4}A_{ki}A_{li} - \dfrac{1}{2}A_{ki}A_l - \dfrac{1}{2}A_{li}A_k + A_lA_k \end{cases}$$

whence

$$B_kB_l + B_lB_k = \frac{1}{2}A_{ki}A_{li} - A_{ki}A_l - A_{li}A_k + (A_kA_l + A_lA_k)$$

and using the relation $(*)$ as above, this gives us

$$B_kB_l + B_lB_k = \frac{1}{2}A_{ki}A_{li} - A_{kl} = B_{ij}$$

Therefore,

$$(C_i)_0 \simeq F[B_{kl} : 0 \leq k <\leq m, \neq i][B_0][B_1, \sigma_1, \delta_1] \dots [B_i] \dots [B_m, \sigma_m, \delta_m]$$

where for all $k < l$ we have

$$\sigma_l(B_k) = B_k \text{ and } \delta_l(B_k) = B_{kl}$$

and trivial actions on the other variables, i.e.

$$(C_i)_0 \simeq Cl_m[A_{1i}, \dots, A_{ii}, \dots, A_{mi}]$$

finishing the proof.

Therefore, $C_i \simeq Cl_m[x_1, \dots, x_m][a_i, a_i^{-1}, \sigma]$. Whereas σ acts trivially on each x_j and B_{kl}, it has a nontrivial action on the B_k, for,

$$\begin{aligned} \sigma(B_k) &= a_i.B_k.a_i^{-1} \\ &= \frac{1}{2}A_{ki} - a_iA_ka_i^{-1} \\ &= \frac{1}{2}A_{ki} + a_ka_i^{-1} - a_{ki}a_{ii}^{-1} = A_k - \frac{1}{2}A_{ki} = -B_k \end{aligned}$$

That is, σ is the main automorphism on Cl_m. So, from the discussion above it follows that

$$X_+(a_{ii}) \simeq \mathrm{Spec}_\sigma(Cl_m[X_1, \ldots, X_m])$$

where Spec_σ denotes the set of all σ-prime ideals. In (a) we have seen that the main automorphism σ on Cl_m permits the elements in a fiber consisting of two elements, i.e.

$$\begin{aligned}\mathrm{Spec}_\sigma(Cl_m[x_1, \ldots, x_m]) &\simeq \mathrm{Spec}(S_m[x_1, \ldots, x_m]) \\ &= A_F^{\frac{m+1}{2}+m}\end{aligned}$$

So, the open set $X_+(a_{ii})$ is homeomorphic to affine space.

Case II : We will write $X_{ij} = za_{ij} + a_{ii} + a_{jj}$ and C_{ij} will be the graded localization of Cl_{m+1} at the central homogeneous multiplicative system

$$\{X_{ij}^n : n \in \mathbb{N}\}$$

Again, C_{ij} contains a homogeneous unit of degree one : $a_i + a_j$. Therefore,

$$C_{ij} = (C_{ij})_0[a_i + a_j, (a_i + a_j)^{-1}, \sigma]$$

where σ is the F-automorphism on $\Gamma_{ij})_0$ given by conjugation in C_{ij} with $a_i + a_j$.

Lemma 3 : $(C_{ij})_0 \simeq Cl_m[X_1, \ldots, x_m]$

Proof.

If we denote $Z_{ij} = a_i + a_j$, then $(C_{ij})_0$ is generated by the elements :

$$\begin{cases} a_{kl}.x_{ij}^{-1} & 0 \le h << m \\ a_k.Z_{ij}^{-1} & k \neq j \end{cases}$$

Let us calculate the commutation rules provided neither k nor $l \neq i$

$$\begin{cases} a_k Z_{ij}^{-1}.a_l Z_{ij}^{-1} = -a_k a_l X_{ij}^{-1} + a_{il} a_k X_{ij}^{-1} Z_{ij}^{-1} + a_{jl} a_k X_{ij}^{-1} Z_{ij}^{-1} \\ a_l Z_{ij}^{-1}.a_k Z_{ij}^{-1} = -a_l a_k X_{ij}^{-1} + a_{ik} a_l X_{ij}^{-1} Z_{ij}^{-1} + a_{jk} a_l X_{ij}^{-1} Z_{ij}^{-1} \end{cases}$$

Therefore, if we denote

$$\begin{aligned} A_k &= a_k Z_{ij}^{-1} && \text{for all} && 0 \le k \neq i, j \le m \\ A_{kl} &= a_{kl}.X_{ij}^{-1} && \text{for} && (k,l) \neq (i,j) \end{aligned}$$

we obtain the commutation relation

$$A_k A_l + A_l A_k = -A_{kl} + (A_{il} + A_{jl}) A_k + (A_{ik} + A_{jk}) A_l \qquad (*)$$

Change the variables in the following way

$$\begin{cases} B_k &= \frac{1}{2}(A_{ik} + A_{jk} - A_k \quad 0 \leq h \neq i, j \leq m \\ B_{kl} &= \frac{1}{2}(A_{ik} + A_{jk})A_{il} + A_{jl} - A_k \quad k \neq i \end{cases}$$

Then we obtain using (*) above

$$B_k B_l + B_l B_k = B_{kl}$$

Now, consider the special case that $k = i$, then

$$\begin{cases} a_i Z_{ij}^{-1} a_l Z_{ij}^{-1} = & -a_i a_l x_{ij}^{-1} + a_{il} a_i X_{ij}^{-1} Z_{ij}^{-1} + a_{jl} a_i X_{ij}^{-1} Z_{ij}^{-1} \\ a_l Z_{ij}^{-1} a_i Z_{ij}^{-1} = & -a_l a_i X_{ij}^{-1} + a_{ii} a_l X_{ij}^{-1} Z_{ij}^{-1} + a_{ij} a_l X_{ij}^{-1} Z_{ij}^{-1} \end{cases}$$

So, if $A_i = a_i Z_{ij}^{-1}$ and $A_{ij} = a_{ij}^{-1}$, the

$$A_1 A_l + A_l A_i = -A_{il} + (A_{il} + A_{jl})A_i + (A_{ii} + A_{ij})A_l$$

And define

$$\begin{cases} B_i &= \frac{1}{2}(A_{ii} + A_{ij}) - A_i \\ B_{il} &= \frac{1}{2}(A_{ii} + A_{ij})(A_{il} + A_{jl}) - A_{il} \end{cases}$$

then

$$B_1 B_l + B_l B_i = B_{il}$$

finishing the proof, since

$$(C_{ij})_0 = F[B_{kl} : 0 \leq k < l \leq m][B_0][B_1, \sigma_1, \delta] \ldots [\check{B}_j] \ldots [B_m . \sigma_m . \delta_m]$$

where for all $k < l$

$$\sigma_l(B_k) = -B_k \text{ and } \delta_l(B_k) = B_{kl}$$

and trivial actions on the other indeterminates.

Now, let us compute the action of σ. If $k \neq i$, then

$$\begin{aligned} \sigma(B_k) &= (a_i + a_j) B_k (a_i + a_j)^{-1} \\ &= \frac{1}{2}(A_{ik} + A_{jk}) + a_k Z_{ij} X_{ij}^{-1} - a_{jk} X_{ij}^{-1} \\ &= A_k - \frac{1}{2}(A_{ik} + A_{jk}) = -B_k \end{aligned}$$

$$\begin{aligned} \sigma(B_i) &= Z_{ij} B_i Z_{ij}^{-1} \\ &= \frac{1}{2}(A_{ii} + A_{ij}) - Z_{ij} A_i Z_{ij}^{-1} \\ &= \frac{1}{2}(A_{ii} + A_{ij}) + A - i Z_{ij} Z_{ij}^{-2} - a_{ii} X_{ij}^{-1} - a_{ij} X_{ij}^{-1} \\ &= -B_i \end{aligned}$$

And as in case I we obtain that

$$X_T(2a_{ij}+a_{ii}+a_{jj}) \simeq \mathrm{Spec}_\sigma(Cl_m[X_1,\dots,X_m]) \\ \simeq A_F^{\binom{m+1}{2}+m}$$

Clearly, the open sets $X_+(a_{ii})$ and $X_+(2a_{ij}+a_{ii}+a_{jj})$ cover the whole of $\mathbb{P}^m_{\mathrm{reg}}$. This completes the proof of.

Theorem 4 : As a topological space, $\mathbb{P}^m_{\mathrm{reg}}$ is homeomorphic to the projective space

$$\mathbb{P}_F^{\binom{m+2}{2}-1}$$

We needed the rather lengthly calculations given above in order to define the structure sheaves later. There is another, easier way of deriving Theorem 4 : intersecting with S_m gives a morphism

$$\begin{array}{ccc} \mathbb{P}^m_{\mathrm{reg}} = \mathrm{Proj}\ Cl_{m+1} & \xrightarrow{\varphi'} & \mathrm{Proj}\ S_{m+1} = \mathbb{P}_F^{\binom{m+2}{2}-1} \\ \downarrow & & \downarrow \\ \mathrm{Spec}\ Cl_{m+1} & \xrightarrow{\varphi} & \mathrm{Spec}\ _{m+1} \end{array}$$

and we need to prove that the fibers $\varphi^{-1}(p)$ of any $p \in \mathrm{Proj}(S_{m+1})$ consist of one element.
Now, $\varphi^{-1}(p)$ is homeomorphic to

$$\mathrm{Spec}(Cl_{m+1} \underset{S_{m+1}}{\otimes} K)$$

where K is the field of fractions of S_{m+1}/p. This algebra is the Clifford algebra associated to the m-ary quadratic form over K

$$\sum \pi(a_{ij})X_iX_j$$

and is therefore isomorphic (as $\mathbb{Z}/2\mathbb{Z}$-graded algebras) to

$$Cl(K,g)\hat{\otimes}\Lambda(W)$$

where g is a regular k-ary quadratic form over K and w is $m-k+1$ dimensional, $k = \mathrm{rank}\ \pi(a_{ij})$. Dividing out the kernel of the augmentation map on $\Lambda(W)$ we have a one-to-one correspondence between $\varphi^{-1}(p)$ and $\mathrm{Spec}Cl(K,q)$. Moreover, the map

$$Cl_{m+1} \to Cl(K,g)$$

is $\mathbb{Z}/z\mathbb{Z}$-graded. If φ^{-1} consists of two elements, then

$$Cl(K,g) \simeq b \oplus b'$$

where b and p' are isomorphic central simple algebras but they are not $\mathbb{Z}/2\mathbb{Z}$-graded. So $\varphi^{-1}(p)$ does not consist out of $\mathbb{Z}/2\mathbb{Z}$-graded prime ideals, so certainly they are not $\mathbb{Z}$-graded, a contradiction. Therefore, φ' is one-to-one. This argument gives also a simplified proof for Theorem III.3.1. of [23].

(c) : Sheaves over $\mathbb{P}^m_{\text{reg}}$

We will now introduce the structure sheaf of a graded module over Cl_{m+1} on $\mathbb{P}^m_{\text{reg}}$. We remind tha reader that F. Van Oystaeyen and A. Verschoren introduced such structure sheaves in [VV]. However, we prefer here to follow a different approach, mainly for two reasons. First, we believe that for p.i.-algebras there is no real need to introduce the machinery of abstract symmetric or bimodule localization theory, but that central localization usually suffices. For one thing, these artificial localizations are allmost never computable except when they agree with central localization, e.g. if the ring is Zariski central or Azumaya. The second, and more fundamental reason, is that their projective schemes are almost never schemes (in the definition of [VV]) except in the trivial cases such as Zariski central rings. A typical example of what might go wrong is $\mathbb{P}^m_{\text{reg}}$. It is possible to find an open cover for it, all opens being homeomorphic to the affine spectrum of an F-algebra but this algebra is allmost never the part of degree zero of the corresponding graded localization since only σ-prime ideals extend.

To remedy this, we associate to an affine p.i. algebra Λ several structure schemes, one for each subring R of the center over which Λ is a finite module, namely

$$(\text{Spec}(R), \underline{\vartheta}_\Lambda) = \underline{\text{Spec}}_R(\Lambda)$$

where $\underline{\vartheta}_\Lambda$ is the usual structure sheaf of the R-module Λ. Similarly, the R-structure sheaf on a left Λ-module M will be

$$(\text{Spec}(R), \underline{\vartheta}_M$$

It is clear that these concepts work only well for affine F-algebras which are finite modules over their center, but since we aim to study only quotients of trace rings of generic matrices this condition is always satisfied. Working with a subring of the center rather that with the center itself provides this theory with some extra (and necessary) flexibility. This approach was motivated by ideas of M. Van den Bergh. If we stick to this framework, one can define a structure sheaf on $\mathbb{P}^m_{\text{reg}}$ such that this ringed space is a scheme. For consider an open set $X_+(a_{ii})$, then this is homeomorphic to the affine spectrum of

$$R_i - \text{Spec } F\left[B_{kl} : 0 \le k \le l \le m\right]\left[A_{1i}, \ldots, \check{A}_{ii}, \ldots, A_{mi}\right] \\ \neq i$$

and $(C_i)_0$ is a finite module over this subring of its center. The structure sheaf $\underline{\vartheta}_{\mathbb{P}^m_{reg}}$ will be defined by

$$\underline{\vartheta}_{\mathbb{P}^m_{reg}} | X_+(a_{ii}) = \underline{\operatorname{Spec}}_{R_i}((C_i)_0)$$

and similarly for the open sets $X_+(2a_{ij}+a_{ii}+a_{jj})$. Then this open is homeomorphic to the affine spectrum of

$$R_{ij} = F\left[B_{kl} : 0 \leq k \leq\leq m\right]\left[A_{1i}, \ldots, \check{A}_{jj}, \ldots, \check{A}_{mi}\right] \\ \neq j$$

and we define

$$\underline{\vartheta}_{\mathbb{P}^m_{reg}} | X_+(2a_{ij} + a_{ii} + a_{pj}) = \operatorname{Spec}\ _{R_{ij}}((C_{ij})_0)$$

It is trivial to verify that $\underline{\vartheta}_{\mathbb{P}^m_{reg}}$ is a sheaf of F-algebras on $\mathbb{P}^m_{reg}$.
Similarly, one can define a structure sheaf $\underline{\theta}_M$ over $\mathbb{P}^m_{reg}$ for any graded left Cl_{m+1}-module M. For take the graded localization M_i of M at the central homogeneous multiplicative set

$$\{T_i^n : n \in \mathbb{N}\}$$

for $T_i \in \{a_{ii}; 2a_{ij} + a_{ii} + a_{jj}\}$. Then M_i is generated by its part of degree zero which is a module over $\Gamma(X_+(Z_i), \underline{\vartheta}_{\mathbb{P}^m_{reg}})$. Then, we define

$$\underline{\vartheta}_M | X_+(Z_i) = (\operatorname{Spec}(R_i), \underline{\vartheta}_{(M_i)_0})$$

In particular we denote for any $n \in \mathbb{Z}$ that the structure sheaf of $Cl_{m+1}(n)$, i.e. the graded module whose part of degree i is equal to $(Cl_{m+1})_{n+i}$, is $\underline{\vartheta}_{\mathbb{P}^m_{reg}}(n)$.
One verifies easily that

$$\bigoplus_{n \in \mathbb{Z}} \Gamma(X_+(Z_i); \underline{\vartheta}_{\mathbb{P}^m_{reg}}(n)) \simeq (Cl_{m+1})_i$$

as $\mathbb{Z}$-graded modules. Now, it is about time to turn attention to $\mathbb{P}^m_{(2)}$.

(d) : Rings of generic trace zero matrices

We have seem above that the trace ring of m generic 2 by 2 matrices is the free polynomial ring

$$\mathbb{T}_{m,2} = \mathbb{T}^0_m[Tr(X_1), \ldots, Tr(X_m)]$$

where $\mathbb{T}^0_m$ is the F-subalgebra generated by the generic trace zero-matrices

$$X_i^0 = X_i - \frac{1}{2} Tr(X_i)$$

For any 2 by 2 matrices A and B having trace zero, one knows that

$$A.B + B.A = Tr(AB)$$

Therefore, sending a_i to X_i^0 and a_{ij} to $\frac{1}{2}Tr(X_i^0 X_j^0)$ we obtain an epimorphism

$$\vartheta_m : Cl_m \to \mathbb{T}_m^0$$

We will now indicate what Ker(φ_m) look like. The center of $\mathbb{T}_m^0$, R_m^0, turns out to be the fixed ring of $F[u_{i1}, u_{i2}, u_{i3} : 1 \leq i \leq m]$ under the canonical action of $SO_3(F)$. It follows from the exact sequence

$$1 \to SO_3(F) \to O_3(f) \to \mathbb{Z}/2\mathbb{Z} \to 1$$

that there is an induced $\mathbb{Z}/2\mathbb{Z}$ action on R_m^0 whose fixed ring is the fixed ring of $F[u_{i1}, u_{i2}, u_{i3}]$ under action of the full orthogonal group $O_3(F)$.
This ring is by classical invariant theory equal to S_m^4, the homogeneous coordinate ring of the variety of all symmetric m by m matrices with entries in F of rank smaller than or equal to 3. Therefore, we obtain the situation

$$\begin{array}{ccccc} Cl_m & \longrightarrow & \mathbb{T}_m^0 & & \\ & & & \nwarrow & \\ \uparrow & & & & R_m^0 \\ & & & \nearrow & \\ S_m & \longrightarrow & S_m^4 & & \end{array}$$

i.e. kernel of φ_m is a (graded) prime ideal of Cl_m lying over the kernel of $\pi_m : S_m \to S_m^4$, i.e. the ideal generated by all 4 by 4 minors of the generic symmetric m by m matrix

$$\mathcal{A} = (a_{ij})_{ij} \in M_m(S_m)$$

Since Ker π_m is a graded prime of S_m, there is only one prime of Cl_m lying over it which is Kerφ_m.

Example :
1. If $m = 2$ or $m = 3$ then φ_M is an isomorphism
2. If $m = 4$, then Kerφ_m is generated by the normalizing element $S_4(a_1, a_2, a_3, a_4)$, i.e. we have an exact sequence

$$0 \to Cl_4.s_4(a_1, a_2, a_3, a_4) \to Cl_4 \to \mathbb{T}_4^0 \to 0$$

(e) Sheaves over $\mathbb{P}_{(2)}^m$

By $\mathbb{P}_{(2)}^m$ we will denote the projective spectrum of the positively graded F-algebra $\mathbb{T}_{m+1}^0$. The epimorphism φ_{m+1} gives a closed immersion

$$\tilde{\varphi}.\mathbb{P}_{(2)}^m \to \mathbb{P}_{reg}^m$$

To economize all our definitions a little we observe that a graded left $\mathbb{T}_{m+1}^0$-module M, is a graded Cl_{m+1}-module, so one has a structure sheaf

$$\underline{\vartheta}_M \text{over } \mathbb{P}_{reg}^m$$

The structure sheaf of M over $\mathbb{P}^m_{(2)}$ is then, of course, defined to be

$$\underline{\vartheta}_M^{(2)} = \tilde{\varphi}^* \underline{\vartheta}_M$$

If it is clear that we are working over $\mathbb{P}^m_{(2)}$ we forget the superscript (2). In particular we denote

$$\underline{\vartheta}_{\mathbb{P}^m_{(2)}} = \underline{\vartheta}^{(2)}_{Cl_{m+1}}$$

$$\underline{\vartheta}_{\mathbb{P}^m_{(2)}}(n) = \underline{\vartheta}^{(2)}_{Cl_{m+2}(n)} \qquad \text{for all } n \in \mathbb{Z}$$

Again, one can cover $\mathbb{P}^m_{(2)}$ by affine open sets namely $X_+(Z_i)$ where

$$Z_i \in \{\varphi_{m+1}(a_{ii}); \varphi_{m+1}(2a_{ij} + a_{ii} + a_{jj})\}$$

and on each such set, one can verify directly

$$\bigoplus_{n \in \mathbb{Z}} \Gamma(X_+(Z_i); \underline{\vartheta}_{\mathbb{P}^m_{(2)}}(n)) \simeq (Cl_{m+1})_{Z_i}$$

as graded modules. where the right hand side denotes the graded localization of Cl_{m+1} at $\{Z_i^n : n \in ?\}$.

We refer the reader to $[L - V - V]$ for a description of the scheme- structure of $\mathbb{P}^m_{(2)}$ on the Azumaya open sets.

$$X_+((X_i^0 X_j^0 - X_j^0 X_i^0)^2)$$

Now, it is about time to calculate some cohomology groups.

2. The cohomology.

In this section we will compute the cohomology of $\mathbb{P}^m_{(2)}$ and $\mathbb{P}^m_{reg}$. Both results can be viewed as a quaternionic generalization of Serre's classical result.

Theorem 5 : (Cohomology of $\mathbb{P}^m_{reg}$)

(1) : $Cl_{m+1} \simeq \bigoplus_{n\in\mathbb{Z}} H^0(\mathbb{P}^m_{reg}, \underline{\vartheta}_{\mathbb{P}^m_{reg}}(n))$

(2) : $H^i(\mathbb{P}^m_{reg}; \underline{\vartheta}_{\mathbb{P}^m_{reg}}(n)) = 0$ for all n and all $0 < i < \binom{m+2}{2} - 1$

(3) : For all $n \in \mathbb{N}$, the dimension of the F-vector space

$$H^{\binom{m+2}{2}-1}(\mathbb{P}^m_{reg} l \underline{\vartheta}_{\mathbb{P}^m_{reg}}(-\binom{m+2}{2} - n))$$

is equal to

$$\sum_{i+2j=n} \binom{m+i}{i} . \binom{\binom{m+1}{2}+j}{j}$$

Proof.

With $\mathcal{F}$ I will denote the quasi-coherent sheaf

$$\bigoplus_{n\in\mathbb{Z}} \underline{\vartheta}_{\mathbb{P}^m_{reg}}(n)$$

Since cohomology commutes with arbitrary direct sums on a Noetherian topological space, the cohomology of $\mathcal{F}$ will be the direct sum of the cohomology of the shaves $\underline{\vartheta}_{\mathbb{P}^m_{reg}}(n)$. Therefore, we aim to compute the cohomology of $\mathcal{F}$ and keep track of the grading by n, so that we can sort out the gradings at the end.
We cover $\mathbb{P}^m_{reg}$ with

$$\mathcal{U} = \{X_+(y_i) : 0 \leq i \leq \binom{m+2}{2} - 1\}$$

where the y_i are the elements from the set

$$\{a_{ii} 1 \leq i \leq m; 2a_{ij} + a_{ii} + a_{jj} : 0 \leq i \leq j \leq m\}$$

For any set of indices $i_i, \ldots, i_p$ between 0 and $\binom{m+2}{2} - 1$ the open set

$$U_{i_0 \ldots i_p} = X_+(y_{i_0}) \cap \ldots \cap X_+(y_{i_p}) = X_+(y_{i_0} \ldots y_{i_p})$$

and by the results of the foregoing section we know that the sections of $\mathcal{F}$ on $U_{i_o,\dots,i_p}$ is equal to the graded localization of Cl_{m+1} at the central homogeneous multiplicative set $\{(y_{i_0}\dots y_{i_p})^n : n \in \mathbb{N}\}$, and the grading by n on $\mathcal{F}$ is the same as the grading of this localization.
Therefore the Čech complex is

$$G^o(\mathcal{U},\mathcal{F}) : 0 \to Cl_{m+1} \to^{\delta_1} \prod_{i_o}(Cl_{m+1})_{y_{i_0}} \to^{\delta_2} \prod_{i_0,\dots,i_1}(cl_{m+1})_{y_{i_0}y_{i_1}}$$
$$\dots \to (Cl_{m+1})_{y_0,\dots,y_\alpha} \to 0$$

where $\alpha = \binom{m+2}{2} - 1$ and all the modules have a natural gradation compatible with that on $\mathcal{F}$, so

$$\bigoplus_{n\in\mathbb{Z}} H^2(\mathbb{P}^m_{reg}, \underline{\vartheta}_{\mathbb{P}^m_{reg}}(n)) \simeq \mathrm{Ker}(\delta_{i+2}/Im(\delta_{i+1})$$

The right hand side of this expression can be computed using local cohomology with respect to the irrelevant ideal $(S_{m+1})_+ = \bigoplus_{i\geq 1}(S_{m+1})_i$. Let us define for every graded left S_{m+1}-module M

$$L(M) = \{m \in M | \exists n > 0 : (S_{m+1})^n_+ .m = 0\}$$

It is well known that L is a left-exact additive functor so we can take the right derived functors R^iL and one has, cfr. e.g. [St]

$$R^{i+1}L(M_{m+1}) = H^{i+1}(Cl_{m+1}) = \mathrm{Ker}\delta_{i+2}/Im\delta_{i+1}$$

and so it will suffice these local cohomology modules. From [St, p. 43-44], we recall that $H^i(M) = 0$ unless $e = \mathrm{depth}(M) \leq i \leq \dim(M) = d$, for $i > 0$, and that $H^e(M) \neq 0, H^d(M) \neq 0$.
Now, Cl_{m+1}, is a graded free module of rank 2^{m+1} over the polynomial subring of its center S_{m+1}, i.e. Cl_{m+1} is a Cohen-Macauley, i.e.

$$\mathrm{depth}(Cl_{m+1} = \dim(Cl_{m+1}) = \binom{m+2}{2}$$

Therefore, if $0 < i < \binom{m+2}{2} - 1$, then

$$\bigoplus_{n\in\mathbb{Z}} H^i(\mathbb{P}^m_{reg}, \underline{\vartheta}_{\mathbb{P}^m_{reg}}(n)) \simeq H^{i+1}(Cl_{m+1}) = 0$$

and clearly,

$$\bigoplus_{n\in\mathbb{Z}} H^0(\mathbb{P}^m_{reg}, \underline{\vartheta}_{\mathbb{P}^m_{reg}}(n)) \simeq H^1(Cl_{m+1}) = 0$$

So, we are left to prove $H^{\binom{m+2}{2}}(Cl_{m+1})$. From [St, Th. 6.4] and Cohen-Macauleyness of Cl_{m+1} we retain that

$$(-1)^{\binom{m+2}{2}}.\mathcal{P}(H^{(m+2)}(Cl_{m+2});t) = \mathcal{P}(Cl_{m+2};t)_\infty$$

where we denote for every graded left Cl_{m+1}-module M its Poincaré series

$$\mathcal{P}(M;t)\sum_{n=0}^{\infty}\dim_F(M_n).t^n.$$

and where $\mathcal{P}(Cl_{m+1};t)_\infty$ signifies that the Poincaré series of Cl_{m+1} is to be expanded as a Laurent series around ∞.
The Poincaré series of Cl_{m+1} is easy to calculate, since as a graded F-vector space, Cl_{m+1} is isomorphic to the commutative polynomial ring

$$F[a_{ij} : 0 \le i < j \le m][a_0, \ldots, a_m]$$

with deg $a_{(ij)} = 2$ and $\deg(a_i) = 1$. Therefore,

$$\mathcal{P}(Cl_{m+1};t) = \frac{1}{(1-t)^{m+1}(1-t^2)^{\binom{m+1}{2}}}$$

and its Laurent series expansion ∞ is equal to

$$(-1)^{\binom{m+2}{2}}.\frac{t^{-\binom{m+2}{2}}}{(1-t^{-1})^{m+1}.(1-t^{-2})^{\binom{m+1}{2}}}$$
$$= (-1)^{\binom{m+2}{2}}.\sum_{n\in\mathbb{N}} t^{-\binom{m+2}{2}-n}.\dim_F(Cl_{m+1})_n$$

Combining all this information, we get that the dimension of

$$H^{\binom{m+2}{2}-1}(\mathbb{P}^m_{reg}, \underline{\vartheta}_{\mathbb{P}^m_{reg}}(-\binom{m+2}{2}-n))$$

is equal to $\dim_F(Cl_m)_n = \sum_{i+2j=n}\binom{m+i}{i}\binom{\binom{m+2}{2}+j}{j}$ finishing the proof.

Pictorially, we get

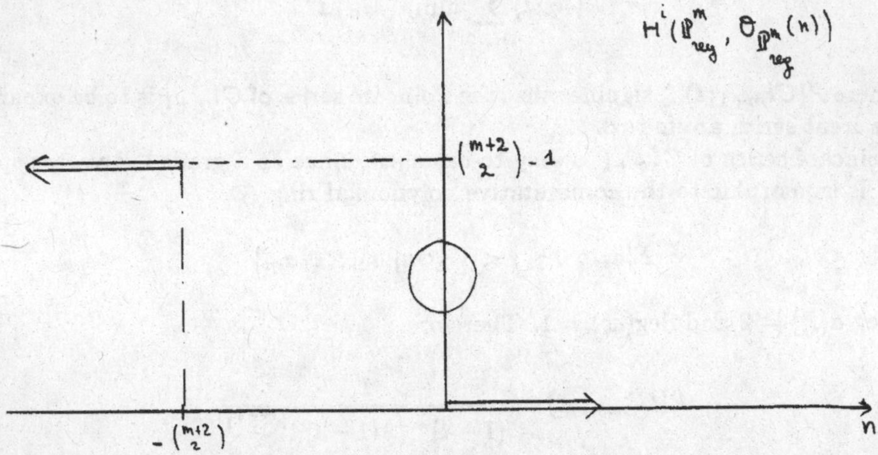

Because we have computed the local structure of $\mathbb{P}^m_{reg}$ explicitly, it might be possible to mimic Serre's proof as in [Ha] to obtain Theorem 5. We have chosen this approach because it clarifies the two main ingredients necessary for a computation of the cohomology of $\mathbb{P}^m_{(2)}$ for which the local structure is not so easy to give.
These two ingredients are :

(A) : a proof that $\mathbb{T}^0_{m+1}$ is a Cohen-Macauley module.

(B) : a functional equation for the Poincaré series of $\mathbb{T}^0_{m+1}$ in order to calculate $\mathcal{P}(\mathbb{T}^0_{m+1}; t)_\infty$.

In my talk before the séminaire in Paris, november '84, I outlined the proof of (A). Since details of this proof will appeare elswhere, e.g. [L3] or [L3], I will sketch here only the main ideas.
First, note that for $m = 2$ of 3, $\mathbb{T}^0_m$ is Cohen-Macauley in view of the isomorphism with Cl_m, mentioned above.
To prove Cohen-Macauleyness for $m \geq 4$, we mimic the argument of Kutz [Ku] to prove Cohen-MacAuleyness of the rings S^h_m, i.e. the homogeneous coordinate ring of symmetric m by m matrices of rank $< k$. He constructs ideals $I_{H,n}$ where

$$H = \{s_0 < s_1 < \ldots < s_l : s_i \in [0, m]\}$$

and $n \in \mathbb{N}$ to be generated by

(0) : the entries of the last s_0 columns of $\mathcal{A} = (a_{ij})$

(1) : The 2 by 2 minors of the last s_1 columns of $\mathcal{A}$

⋮

(l) : The $l+1$ by $l+1$ minors matrices of the last s_l columns of $\mathcal{A}$

(*) : The entries of the last n columns of the first row of $\mathcal{A}$

Remark that the ideal generated by all k by k minors (i.e. $\mathrm{Ker}(S_m \to S_m^k)$) is of the form with

$$H = \{0 < 1 < 2 \ldots < k-2 < m\}; n = 0$$

Kutz shows that in case $n = s_p$ for some p, then $I_{H,n}$ is a prime ideal and if $s_p < n < s_{p+1}$, then

$$I_{H,n} = I_{H',n} \cap I_{H,n'}$$

where $n' = s_{p+1}$ and $H' = \{s_o < \ldots < s_{p-1} < n < s_{p+1} \ldots\}$ where both $I_{I',n}$ and $I_{H,[p'}$ are (graded) prime ideals of S_m.
There is a unique prime ideal $J_{H,n}$ of Cl_m lying over such a prime and we define for $s_p < n < s_{p+1}$

$$J_{H,n} = J_{H',n} \cap J_{H,p'}$$

Using some structure theory of Clifford algebras one can show that if $n = s_p + 1$, then

$$J_{H',p'} = J_{H',n} + J_{H,p'}$$

and if $n = s_p$, then

$$J_{H,n+1} = J_{H,n} + Cl_m.a_{1,m-n}$$

Now, using these two relations one can show by an induction argument that

$$pd_{Cl_m}(Cl_m/J_{H,n}) = \mathrm{Kdim}(Cl_m) - \mathrm{Kdim}(Cl_m/J_{H,n})$$

Whenever $n = s_p$ or $n = s_p + 1$ for some p.
In particular

$$\begin{aligned} pd_{Cl_m}(\mathbb{T}_m^0) &= \mathrm{Kdim}(Cl_m) - \mathrm{Kdim}(\mathbb{T}_m^0) \\ &= \binom{m+1}{2} - 3m + 3 \\ &= \frac{(m-2)(m-3)}{2} \end{aligned}$$

Finally, using that Cl_m is a free module of finite rank over S_m and that Cl_m is a maximal order having trivial normalizing class group one can show that

$$\begin{aligned} \mathrm{Ext}^i_{S_m}(\mathbb{T}_m^0, S_m) &\simeq \mathrm{Ext}^i_{Cl_m}(\mathbb{T}_m^0, \mathrm{Hom}_{S_m}(Cl_m, S_m)) \\ &\simeq \mathrm{Ext}^i_{Clm}(\mathbb{T}_m^0, Cl_m) \end{aligned}$$

and so $pd_{S_m}(\mathbb{T}_m^0) = \dim S_m - \dim \mathbb{T}_m^0$ whence $\mathbb{T}_m^0$ is a Cohen-Macauley module over S_m. For more details, see [L3].

Now, let us consider objective (B). The map

$$Tr(-X_{m+1})\mathbb{T}_{m,2} \to \mathcal{R}_{m+1,2}$$

is a linear injection onto the subspace of elements which are homogeneous of degree one in X_{m+1}. In terms of multilinear Poincaré series this means that

$$\mathcal{P}(\mathbb{T}_{m,2};t_1,\ldots,t_m) = \frac{\partial}{\partial t_{m+1}}\mathcal{P}(\_m+1,2;t_1,\ldots,t_{m+1})\Big|_{t_{m+1}=0}$$

Now, $\mathcal{R}_{m+1,2} = \mathcal{R}^0_{m+1}[Tr(X_1),\ldots,Tr(X_{m+1})]$ where $\mathcal{R}^0_{m+1}$ is the ring of invariants under action of $SO_3(F)$ whose multilinear Poincaré series were calculated by e.g. H. Weyl.
Combining all this, we gave a rational expression for the Poincaré series of $\mathbb{T}_{m,2}$ in [1].

Proposition 6 [L1]

The Poincaré series of the trace ring of m generic 2 by 2 matrices has the following rational expression

$$\mathcal{P}(\mathbb{T}_{m,2};t_1,\ldots,t_m) = \frac{e_m\Delta_1 - (e_m + e_1e_m + e_{m-1})\Delta_2}{e_m^2\prod_{j=1}^m(1-tj)\prod_{i<k}^m(T_k-t_i)\prod_{i\leq k}^m(1-t_it_k)}$$

Here, the e_i are the i-th elementary symmetric function in m variables and the Δ_i are defined by

$$\Delta_1 = \det\begin{bmatrix} 1+t_1^{2m-1} & \ldots & 1+t_m^{2m-1} \\ t_1^2+t_1^{2m-3} & \ldots & t_m^2+t_m^{2m-3} \\ t_1^3+t_1^{2m-4} & \ldots & t_m^3+t_m^{2m-n} \\ \vdots & & \vdots \\ t_1^{m-2}+t_1^{M+1} & \ldots & t_m^{m-2}+t_m^{m+1} \\ t^{m-1} & \ldots & t_m^{m-1} \\ t_1^m & \ldots & t_m^m \end{bmatrix}$$

$$\Delta_2 = \det\begin{bmatrix} t_1+t_1^{2m-2} & \ldots & t_m+t_m^{2m-2} \\ t_1^2+t_1^{2m-3} & \ldots & t_m^2+t_m^{2m-3} \\ \vdots & & \vdots \\ t_1^{m-2}+t_1^{m+2} & \ldots & t_m^{m-2}+t_m^{m+2} \\ t_1^{m-1} & \ldots & t_m^{m-1} \\ t_1^m & \ldots & t_m^m \end{bmatrix}$$

and as an immediate consequence of this, one obtains :

Corollary 7 [L1] (Functional equation)

The Poincaré series of the trace ring of m-generic 2 by 2 matrices satisfies the functional equation

$$\mathcal{P}(\mathbb{T}_{m,2}; \frac{1}{t} = -t^{4m}.\mathcal{P}(\mathbb{T}_{m,2}; t)$$

Another, more ringtheoretocal proof of this fact using Cohen-Macauleybess of $\mathbb{T}_{m,2}$ and that $\mathbb{T}_{m,2}$ is a maximal order having trivial normalizing classgroups can be found in [L3].
We have now all material at our disposal to calculate the cohomology of $\mathbb{P}^m_{(2)}$.

Theorem 8 : (Cohomology of $\mathbb{P}^m_{(2)}$)

(1) : $\mathbb{T}^0_{m+1} \simeq \bigoplus_{n\in\mathbb{Z}} H^0(\mathbb{P}^m_{(2)}, \underline{\vartheta}_{\mathbb{P}^m_{(2)}}(n))$

(2) : $H^i(\mathbb{P}^m_{(2)}, \underline{\vartheta}_{\mathbb{P}^m_{(2)}}(n)) = 0$ for all n and $0 < i < 3m-1$

(3) : There is a one-to-one correspondence between a basis of the F-vectorspace

$$H^{3m-1}(\mathbb{P}^m_{(2)}, \underline{\vartheta}_{\mathbb{P}^m_{(2)}}(-3m-3-n))$$

and standard young tableaux of length ≤ 3 filled with entries from 0 to m having n boxes.

Proof.

Of course we take as an affine cover of $\mathbb{P}^m_{(1)}$

$$Tl = \{y_i : 0 \leq i \leq \binom{m+2}{2} - 1\}$$

where the y_i are the elements from the set

$$\{\phi(a_{ii}) : 0 \leq i \leq m; \phi(2a_{ij} + a_{ii} + a_{jj}) : 0 \leq i < j \leq m\}$$

where $\phi : Cl_{m+2} \to \mathbb{T}^0_{m+1}$ is the natural epimorphism. As in the proof of Theorem 5 we consider the quasi-coherent sheaf

$$\bigoplus_{n\in\mathbb{Z}} \underline{\vartheta}_{\mathbb{P}^m_{(2)}}(n)$$

and similarly one deduces that

$$\bigoplus_{n\in\mathbb{Z}} H^i(\mathbb{P}^m_{(2)}, \underline{\vartheta}_{\mathbb{P}^m_{(2)}}(n)) \simeq H^{i+1}(\mathbb{T}^0_{m+1})$$

We know that $\mathbb{T}^0_{m+1}$ is a Cohen-Macauley module module over S_{m+1} of dimension $3(m+1) - 3 = 3m$, so part (2) of the theorem is proved.

As for part (3), we know that

$$\mathcal{P}(H^{3m}(\mathbb{T}^0_{m+1});t) = \mathcal{P}(\mathbb{T}^0_{m+1};t)_\infty.(-1)^{3m}$$

From the functional equation for the Poincaré series of $\mathbb{T}_{m+1,2}$ and the fact that $\mathbb{T}_{m+1,2} = \mathbb{T}^0_{m+1}[Tr(X_0),\ldots,Tr(X_m)]$ i.e.

$$\mathcal{P}(\mathbb{T}_{m+1,2};t) = \frac{1}{(1-t)^{m+1}}.\mathcal{P}(\mathbb{T}^0_{m+1};t)$$

we find that

$$\mathcal{P}(\mathbb{T}^e_{m+1};\frac{1}{t}) = (-1)^{3m}t^{3m+3}\mathcal{P}(\mathbb{T}^0_{m+1};t)$$

i.e.

$$\mathcal{P}(\mathbb{T}^0_{m+1};t)_\infty = (-1)^{3m}t^{-3m-3}.\mathcal{P}(\mathbb{T}^0_{m+1};\frac{1}{t})$$

and therefore

$$\mathcal{P}(H^{3m}(\mathbb{T}^0_{m+1});t = \sum_{n\in\mathbb{N}} t^{-3m-3-n}.\dim_F(\mathbb{T}^0_{m+1})_n$$

and from the work of C. Procesi, [P2], we retain that there isa a one-to-one correspondence between standard Young tableaux of shape $\sigma = 3^a2^b1^c$ for $a,b,c,\in\mathbb{N}$. i.e. a diagram consisting of a rows of length 3, b rows of length 2 and c rows of length 1

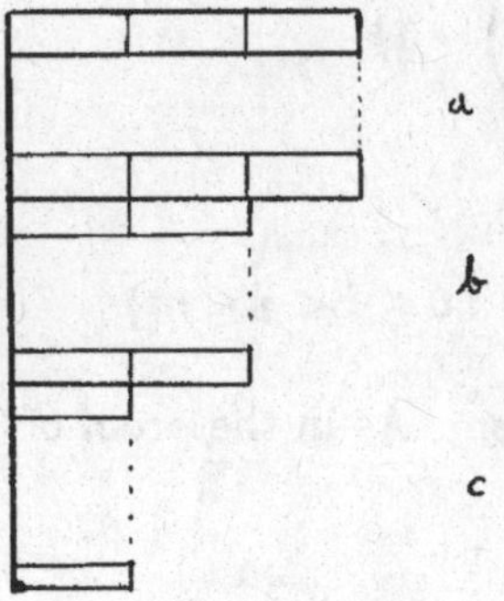

filled with indices from 0 tot m such that the numbers in every row strictly increase and that the numbers in every column do not decrease; and an F-vector space basis of $\mathbb{T}^0_{m+1}$. Moreover, the degree of an element corresponding to a standard Young tableau of shape $\sigma = 3^a2^b1^c$ is equal to the number of boxes in the diagram, i.e. is $3a+2b+c$.
This finishes the proof of the theorem.

Pictorially, we have the situation :

From this point on, it is easy to derive as in the commutative case a Grothendieck-Serre duality result for coherent sheaves over $\mathbb{P}^m_{(2)}$ or $\mathbb{P}^m_{(reg)}$. We leave this as an (easy) exercise to the reader, cfr. [A-K] for the commutative proof.

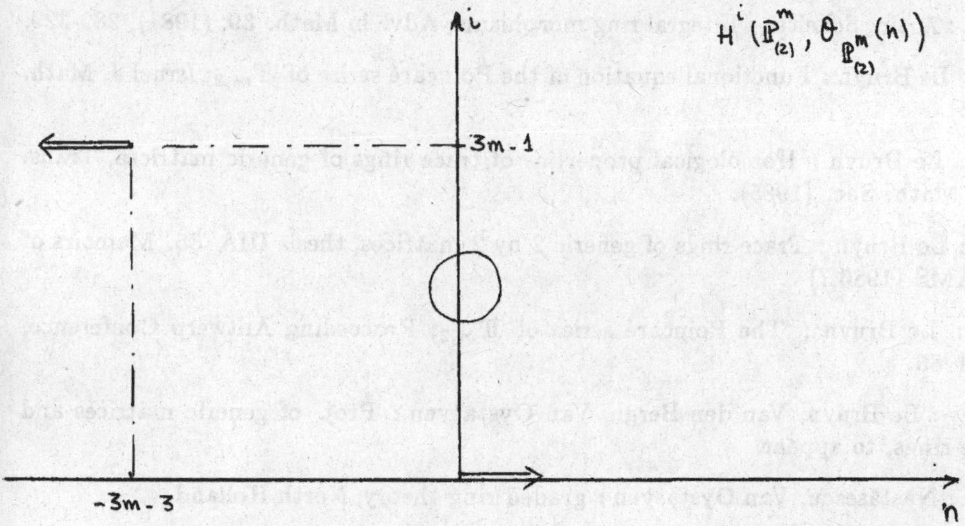

References :

[A-K] : Altman, Kleiman : Introduction to Grothendieck duality, Springer Verlag LNM.

[Ar] : Artin : Azumaya algebras and finite dimensional representations of rings. J. Alg. II (1969), 532-563.

[A-S] : Artin, Schelter : Integral ring morphisms, Adv. in Math. 39, (1981), 289-329.

[L1] : Le Bruyn : Functional equation of the Poincaré series of $\mathbb{T}_{m,2}$; Israel J. Math. (1985).

[L2] : Le Bruyn : Homological properties of trace rings of generic matrices, Trans. Am. Math. Soc. (1985).

[L3] : Le Bruyn : Trace rings of generic 2 by 2 matrices, thesis UIA '85. Memoirs of the AMS (1986 ?)

[L4] : Le Bruyn : The Poincaré series of $\mathbb{T}_{m,2}$; Proceeding Antwerp Conference, April '85.

[LVV] : Le Bruyn, Van den Bergh, Van Oystaeyen : Proj. of generic matrices and trace rings, to appear.

[NV] : Năstăsescu, Van Oystaeyen : graded ring theory, North Holland.

[St] : Stanley : Combinatorial methods in commutative ring theory, Birkhäuser.

[P1] : Procesi : Finite dimensional representations of algebra, Israel J. Math. 19, (1974), 169-184.

[P2] : Procesi : Computing with 2 by 2 matrices, J of Algebra 87, (1984), 342-359.

[Lam] : Lam : Algebraic theory of quadratric forms, Benjamin, (1973).

[VV] : Van Oystaeyen, Verschoren : Noncommutative algebraic geometry. Springer LNM 887.

[Ha] : Hartshorne : Algebraic geometry, Springer Verlag

[Ku] : Kutz : Cohen-Macauley rings and ideal theory in rings of invariants of algebraic groups, Trans. Am. Math. Soc. 197, (1974), 115-129.

DIFFERENTIAL OPERATORS ON THE AFFINE AND PROJECTIVE LINES IN CHARACTERISTIC $p > 0$

by

S.P. SMITH

Let k be a field, and denote by $\mathbb{A}^1$ (or $\mathbb{A}^1_k$) and $\mathbb{P}^1$ (or $\mathbb{P}^1_k$) the affine and projective lines over k. When k is of characteristic 0 the rings of differential operators on $\mathbb{A}^1$ and $\mathbb{P}^1$ (which we denote $D(\mathbb{A}^1_o)$ and $D(\mathbb{P}^1_o)$) have been extensively studied, and are considered to be well understood. In contrast, if char $k = p > 0$, the rings of differential operators on $\mathbb{A}^1$ and $\mathbb{P}^1$ (which we denote $D(\mathbb{A}^1_p)$ and $D(\mathbb{P}^1_p)$) have not been studied at all. The purpose of this note is to begin an investigation into $D(\mathbb{A}^1_p)$ and $D(\mathbb{P}^1_p)$.

Before we outline some of our results, we give a brief account of the wider context in which $D(\mathbb{A}^1_o)$ and $D(\mathbb{P}^1_o)$ appear (and which accounts for their significance). First, if one is to study differential operators on any affine or projective variety then $D(\mathbb{A}^1)$ and $D(\mathbb{P}^1)$ are the first cases to examine. However, another important motivation is the connection of $D(\mathbb{A}^1_o)$ and $D(\mathbb{P}^1_o)$ with the representation theory of finite dimensional Lie algebras in characteristic zero. The recent history of $D(\mathbb{A}^1_o)$ (known as the Weyl algebra) begins with Dixmier's papers [3] and [4]. He showed that if $\underset{\sim}{g}$ is a finite dimensional nilpotent Lie algebra over $\mathbb{C}$, then the primitive factor rings of $U(\underset{\sim}{g})$, the enveloping algebra of $\underset{\sim}{g}$, are of the form $D(\mathbb{A}^n_{\mathbb{C}}) \cong D(\mathbb{A}^1_{\mathbb{C}}) \otimes_{\mathbb{C}} \cdots \otimes_{\mathbb{C}} D(\mathbb{A}^1_{\mathbb{C}})$. Hence, the irreducible representations of $\underset{\sim}{g}$ are precisely the simple modules over $D(\mathbb{A}^n_{\mathbb{C}})$ for various n. For example, if $\underset{\sim}{g}$ is the 3-dimensional Heisenberg Lie algebra then the infinite dimensional irreducible representations of $\underset{\sim}{g}$ are precisely the simple modules over $D(\mathbb{A}^1_{\mathbb{C}})$.

The ring $D(\mathbb{P}^1_{\mathbb{C}})$ arises in a similar way. Let G be a connected complex semi-simple Lie group with Borel subgroup B; then G/B is a complex projective algebraic variety ($\mathbb{P}^1_{\mathbb{C}}$ arises as SL(2)/B), and the ring of global regular differential operators on G/B, D(G/B), is isomorphic to a primitive factor ring of $U(\underset{\sim}{g})$ where $\underset{\sim}{g}$ is the Lie algebra of G. See [1] where this idea is exploited to verify the

Kazhdan-Lusztig conjectures on Verma modules.

The corresponding connections between representations of characteristic p Lie algebras and modules over $D(\mathbb{A}^1_p)$ and $D(\mathbb{P}^1_p)$ are not studied here. Rather, we concern ourselves with the ring theoretic properties of $D(\mathbb{A}^1_p)$ and $D(\mathbb{P}^1_p)$ and examine to what extent their structure parallels or diverges from $D(\mathbb{A}^1_0)$ and $D(\mathbb{P}^1_0)$. It is largely a matter of taking a result in characteristic zero and asking whether the same result holds in characteristic p, and if not, in what sense is it false.

In Table 1, below, the properties of $D(\mathbb{A}^1)$ in characteristics zero and p are set out side by side. Let us mention just a few of them. $D(\mathbb{A}^1_0)$ is finitely generated and Noetherian - both these are false for $D(\mathbb{A}^1_p)$. Much of the "bad" behaviour of $D(\mathbb{A}^1_p)$ can be attributed to the lack of some sort of finiteness condition (in particular, the question of whether every endomorphism of a simple $D(\mathbb{A}^1_p)$-module is algebraic over k, is difficult because one has no finiteness condition which might allow a result concerning generic flatness of the associated graded algebra to be established). For a similar reason Gelfand-Kirillov dimension, which is an effective tool for $D(\mathbb{A}^1_0)$, does not seem to be useful for $D(\mathbb{A}^1_p)$. But, all is not lost. For example, if k[t] denotes the co-ordinate ring of $\mathbb{A}^1$, and if $0 \neq f \in k[t]$ then $k[t,f^{-1}]$ is a $D(\mathbb{A}^1)$-module. In characteristic zero, $k[t,f^{-1}]$ is an Artinian module, and the usual proof involves Gelfand-Kirillov dimension. Nevertheless, in characteristic p, $k[t,f^{-1}]$ is also an Artinian $D(\mathbb{A}^1_p)$-module, and the proof makes use of one structural feature of $D(\mathbb{A}^1_p)$ that has no analogue in $D(\mathbb{A}^1_0)$. Namely that $D(\mathbb{A}^1_p) = \bigcup_{n=0}^{\infty} \mathrm{End}_{k[t^{p^n}]} k[t]$, is a union of matrix algebras over commutative rings (whereas $D(\mathbb{A}^1_0)$ is a domain). One question which appears in [3] and remains unanswered to date, is whether $D(\mathbb{A}^1_0)$ has a proper subring isomorphic to $D(\mathbb{A}^1_0)$. It is quite easy to construct a proper subring of $D(\mathbb{A}^1_p)$ which is isomorphic to $D(\mathbb{A}^1_p)$.

Although $D(\mathbb{P}^1_{\mathbb{C}})$ is a primitive factor ring of $U(sl(2,\mathbb{C}))$, the natural map from $\mathrm{Hyp}(sl(2,k))$, the hyperalgebra of $sl(2,k)$, to $D(\mathbb{P}_k)$ is not surjective if char $k = p > 0$.

$D(\mathbb{P}^1_0)$ has a unique two sided ideal (apart from 0 and $D(\mathbb{P}^1_0)$) and this ideal is of codimension 1; the analogous statement for $D(\mathbb{P}^1_p)$ is also true. Whereas $K_0(D(\mathbb{P}^1_0)) = \mathbb{Z} \oplus \mathbb{Z}$, $K_0(D(\mathbb{P}^1_p)) = \mathbb{Z} \oplus \mathbb{Z}[1/p]$; the lattice of order ideals in $K_0(D(\mathbb{P}^1_p))$ is isomorphic to the lattice of two sided ideals in $D(\mathbb{P}^1_p)$.

TABLE 1

Properties of $D(\mathbb{A}^1_k)$

Characteristic zero	Characteristic $p > 0$
finitely generated	not finitely generated
Noetherian	not Noetherian
simple ring	simple ring
domain	not a domain
gl.dim. = 1	gl.dim. = 1
K.dim. = 1	K.dim. does not exist
GK.dim. = 2	GK. dim. = 1.
centre = k	centre = k
$K_0 = \mathbb{Z}$	$K_0 = \mathbb{Z}[1/p]$
Every derivation is inner	There exists a non-inner derivation
If I is a left ideal with $I \cap k[t] \neq 0$ and $I \cap k[d/dt] \neq 0$, then $I = D(\mathbb{A}^1)$	If char k = 2 then $Dt + Dx_1 \neq D(\mathbb{A}^1)$
If $0 \neq f \in k[t]$ then $k[t,f^{-1}]$ is Artinian	If $0 \neq f \in k[t]$ then $k[t,f^{-1}]$ is of finite length
k[t] is a simple module	k[t] is a simple module
D/Dt is a simple module	D/Dt is a simple module
Open question whether $D(\mathbb{A}^1)$ has a proper subalgebra isomorphic to $D(\mathbb{A}^1)$	$D(\mathbb{A}^1)$ contains a proper subalgebra isomorphic to $D(\mathbb{A}^1)$ viz $k[t^p, x_p, x_{2p}, x_{3p}, \ldots]$
If M is a simple module $\mathrm{End}_D M$ is algebraic over k	Not known

My initial interest in these ideas was aroused during conversations and correspondence with Ken Goodearl. I am indebted to him for his generous comments and assistance, especially relating to matters concerning K-theory. My thanks also go to C.R. Hajarnavis for many useful conversations during the preparation of these notes.

§1. DIFFERENTIAL OPERATORS

Let k be any commutative ring, and A any commutative k-algebra. Then $\mathrm{End}_k A$ may be made into an $A \otimes_k A$-module by defining $((a \otimes b)\theta)(c) = a\theta(bc)$ for $\theta \in \mathrm{End}_k A$ and $a,b,c \in A$. We write $[a,\theta]$ for $(a \otimes 1 - 1 \otimes a)\theta$, so $[a,\theta](b) = a\theta(b) - \theta(ab)$.

DEFINITION 1.1 The space of k-*linear differential operators of order* $\leq n$ *on* A, $\mathrm{Diff}_k^n A$, is defined inductively by $\mathrm{Diff}_k^{-1} A = 0$, and for $n \geq 0$,
$\mathrm{Diff}_k^n A = \{\theta \in \mathrm{End}_k A \mid [a,\theta] \in \mathrm{Diff}_k^{n-1} A$ for all $a \in A\}$. The *ring of* k-*linear differential operators on* A is $D(A) = \bigcup_{n=0}^{\infty} \mathrm{Diff}_k^n A$. If X is an affine algebraic variety over the field k with ring of regular functions A, we write $D(X) = D(A)$.

REMARK 1.2 (1) $\mathrm{Diff}_k^n A$ is an $A \otimes A$-submodule of $\mathrm{End}_k A$

(2) If $\theta \in \mathrm{End}_k A$, then $\theta \in \mathrm{Diff}_k^n A$, if and only if,

$$[a_0[a_1 \ldots [a_n,\theta]\ldots]] = 0 \text{ for all } a_0,a_1,\ldots,a_n \in A.$$

(3) We refer the reader to [10] for a more comprehensive introduction to rings of differential operators on commutative rings.

(4) It is an easy exercise to verify that if k is a field of characteristic zero, and k[t] is the ring of regular functions on $\mathbb{A}_k^1$, then $D(\mathbb{A}_k^1) = k[t,d/dt]$ where d/dt is the usual differentiation operator acting on the polynomial ring k[t]. As elements of $\mathrm{End}_k k[t]$ one has $(d/dt)t - t(d/dt) = 1$.

DEFINITION 1.3 Denote by $\mu: A \otimes_k A \to A$ the multiplication map $\mu(a \otimes b) = ab$. This is a k-algebra map (also an A-module map for either the right or left A-module structure on $A \otimes_k A$). Put $I = \ker \mu$.

THEOREM 1.4 (Heynemann-Sweedler [9], Grothendieck [8]). *Let* $\theta \in \mathrm{End}_k A$. *Then* $\theta \in \mathrm{Diff}_k^n A$, *if and only if*, $I^{n+1}.\theta = 0$.

§2. PROPERTIES OF $D(\mathbb{A}_p^1)$

Write $D = D(\mathbb{A}_p^1)$, and consider D as the ring of k-linear differential operators on k[t], the polynomial ring in t, over the field k of characteristic $p > 0$.

The following result was arrived at during conversation and correspondence with

Ken Goodearl, and I am grateful for his allowing me to include it here.

PROPOSITION 2.1 $D = \bigcup_{n=0}^{\infty} \mathrm{End}_{k[t^{p^n}]}k[t]$ *and* $\mathrm{Diff}_k^{p^n-1} k[t] = \mathrm{End}_{k[t^{p^n}]}k[t]$.

Proof Let $\theta \in \mathrm{End}_k k[t]$. Notice that $I = \ker(\mu: k[t] \otimes_k k[t] \to k[t])$ is generated as an ideal by $1 \otimes t - t \otimes 1$. Hence I^{p^n} is generated by $(1 \otimes t - t \otimes 1)^{p^n} = 1 \otimes t^{p^n} - t^{p^n} \otimes 1$. So $\theta \in \mathrm{Diff}_k^{p^n-1} k[t]$, if and only if, $I^{p^n}.\theta = 0$. That is, if and only if, $0 = (1 \otimes t^{p^n} - t^{p^n} \otimes 1).\theta = \theta t^{p^n} - t^{p^n}\theta$. So θ is a differential operator of order $\leq p^n-1$, if and only if $\theta \in \mathrm{End}_{k[t^{p^n}]}k[t]$. This proves the result. □

We shall write $D_n = \mathrm{Diff}_k^{p^n-1} k[t]$. So we have just shown that $D_n \cong M_{p^n}(k[t^{p^n}])$, the $p^n \times p^n$ matrix ring over $k[t^{p^n}]$.

COROLLARY 2.2 (1) D *is not a finitely generated* k*-algebra;*

(2) D *does not contain any primitive idempotents; in fact if* $0 \neq e \in D$ *is idempotent then there exists a set of* p *mutually orthogonal idempotents* $e_1, \ldots, e_p$ *such that* $e = e_1 + \ldots + e_p$;

(3) D *contains an infinite direct sum of non-zero left ideals;*

(4) D *is not Noetherian;*

(5) D *does not have Krull dimension (in the sense of Gabriel and Rentschler).*

Proof (3), (4), (5) are immediate consequences of (2), and (1) is obvious, since any finite set of elements of D lies in some D_n, and so can at best generate D_n which is a proper subalgebra of D.

To prove (2), let $0 \neq e \in D$ be an idempotent. Suppose $e \in D_n = \mathrm{End}_{k[t^{p^n}]}k[t]$. Write $k[t] = U \oplus V$, a direct sum of $k[t^{p^n}]$-submodules, where $e|_U = \mathrm{Id}|_U$ and $e(V) = 0$. As $e \neq 0$, U is non-zero, and as a $k[t^{p^{n+1}}]$-module, $U = U_1 \oplus \ldots \oplus U_p$ is a direct sum of p non-zero $k[t^{p^{n+1}}]$-modules. Now $e = e_1 + \ldots + e_p$ where e_j is the projection of k[t] onto U_j with kernel $V \oplus U_1 \oplus \ldots \oplus \hat{U}_j \oplus \ldots \oplus U_p$ (omit U_j from the sum). One checks that each e_j is a $k[t^{p^{n+1}}]$-module map, hence an element of D_{n+1}, and that the e_j are mutually orthogonal idempotents. □

A concrete illustration of (2) above, is the following: if $e_n: k[t] \to k[t]$ is the

$k[t^{p^n}]$-linear map defined by $e_n(t^i) = \delta_{i,p^n-1}t^i$ for $0 \le i \le p^n$, then $\{e_1, e_2, \ldots\}$ is an infinite set of mutually orthogonal idempotents.

PROPOSITION 2.3 $K_0(D) \cong \mathbb{Z}[1/p]$

Proof $D_n \cong M_{p^n}(k[t^{p^n}])$ and one has that $K_0(D_n) = K_0(k[t^{p^n}])$ (as K_0 is defined in terms of the category of modules over D_n) and it is known that $K_0(k[t^{p^n}]) = \mathbb{Z}$. The inclusions $D_1 \to D_2 \to D_3 \to \ldots$ induce maps on the K_0 groups $\mathbb{Z} \xrightarrow{p} \mathbb{Z} \xrightarrow{p} \mathbb{Z} \to \ldots$. The maps are multiplication by p. As K_0 commutes with direct limits [7] we get $K_0(D) = \mathbb{Z}[1/p]$. □

An order unit is $1 = [R]$, and the order relation is the usual order relation on $\mathbb{Z}[1/p]$.

PROPOSITION 2.4 *Not every derivation of* D *is inner.*

Proof Define $\Delta: D \to D$ by $\Delta(d) = [t + t^p + t^{p^2} + \ldots, d]$. This actually makes sense: for $n \gg 0$, $d \in D_{n+1}$ and so d commutes with $t^{p^{n+1}}$, and hence with t^{p^m} for all $m > n$; therefore $\Delta(d) = [t + t^p + \ldots t^{p^n}, d]$ for $d \in D_{n+1}$.

Suppose Δ is inner, say $\Delta = \mathrm{ad}(y)$ for some $y \in D$. Let $y \in D_n$. As $\Delta(t) = 0$, y commutes with $k[t]$, hence $y \in k[t]$. For all n we have $\Delta - \mathrm{ad}y|_{D_{n+1}} = 0$ but we have just seen that $\Delta|_{D_{n+1}} = \mathrm{ad}(t + t + \ldots + t^{p^n})$. Hence $\mathrm{ad}(t+t^p+\ldots+t^{p^n}-y)|_{D_{n+1}} = 0$, and so $t + t^p + \ldots t^{p^n} - y$ belongs to the centre of D_{n+1} $(= k[t^{p^{n+1}}])$ for all n; this is impossible. □

PROPOSITION 2.5 Centre (D) = k.

Proof Centre $(D_n) = k[t^{p^n}]$ and $\bigcap_{n=0}^{\infty} k[t^{p^n}] = k$. The proposition is an immediate consequence. □

Another description of D is also useful. For each $i \in \mathbb{N}$, let x_i be the k-linear map on $k[t]$ given by $x_i(t^m) = \binom{m}{i}t^{m-i}$ where the binomial coefficient $\binom{m}{i}$ is evaluated (mod p). One should think of x_i as acting like $(1/i!)\partial^i/\partial t^i$; even though $1/i!$ does not make sense in k if $i \ge p$, this analogy can be made rigorous, as in Theorem 2.7 below. The analogy is useful in noticing relationships such as $x_i x_j = \binom{i+j}{i}x_{i+j}$.

THEOREM 2.6 $D_n = k[t,x_1,x_2,\ldots,x_{p^n-1}]$ *and* $D = k[t,x_1,x_2,\ldots]$.

Proof To see that x_m is a differential operator of order $\le m$, notice that $x_o = 1 \in D_o$ and $[x_m,t] = x_{m-1}$ then use the inductive Definition 1.1. Thus $k[t,x_1,\ldots,x_{p^n-1}] \subset D_n$.

Viewing $D_n \cong M_{p^n}(k[t^{p^n}])$, there is a basis for D_n as a $k[t^{p^n}]$-module given by the maps $\theta_{ij}:k[t] \to k[t]$ for $0 \le i, j < p^n$ where θ_{ij} is the $k[t^{p^n}]$-module map defined by $\theta_{ij}(t^m) = \delta_{jm}t^{m+i-j}$ for $0 \le m < p^n$. The θ_{ij} are just the matrix units (for the basis $1,t,\ldots,t^{p^n-1}$ of $k[t]$ as a $k[t^{p^n}]$-module).

One computes that $\theta_{ij} = t^i x_{p^n-1} t^{p^n-1-j}$ (the point being that $\binom{\ell}{p^n-1}$ is zero for all $\ell \in \mathbb{N}$ unless $\ell = p^n-1$). Thus $\theta_{ij} \in k[t,x_1,\ldots,x_{p^n-1}]$. This completes the proof. □

Recall that $D(\mathbb{Q}[t]) = \mathbb{Q}[t,\partial/\partial t]$. One can easily check that the $\mathbb{Z}$-module spanned by all elements of the form $t^j(1/i!)\partial^i/\partial t^i$ is in fact a $\mathbb{Z}$-subalgebra; write $S=\mathbb{Z}[t,\partial/\partial t,(1/2!)\partial^2/\partial t^2,\ldots]$. Of course $S = D(\mathbb{Z}[t])$, the ring of $\mathbb{Z}$-linear differential operators on $\mathbb{Z}[t]$. The following is straightforward.

THEOREM 2.7 $D(k[t]) \cong k \otimes_{\mathbb{Z}} S$ *where the isomorphism is given by* $x_i \to 1 \otimes (1/i!)\partial^i/\partial_t^{\,i}$ *and* $t \to t$.

The proof that D is a simple ring is inevitably a little more complicated than the proof in the characteristic zero case - if one recalls the characteristic zero proof, one part of it is the observation that if I is a non-zero ideal and $0 \ne a \in I$ then for some n, $\mathrm{ad}^n(\partial/\partial t)(a) \in k[\partial/\partial t]\backslash 0$, so there exists $0 \ne b \in I$ with $b \in k[\partial/\partial_t]$ and then for some m $\mathrm{ad}^m(t)(b) \in k\backslash\{0\}$, so I contains a scalar. However, if char $k = 2$, $\mathrm{ad}(\partial/\partial t)(t^2) = 0$.

Hence we require the following technical result.

LEMMA 2.8 $[x_\ell, t^m] = \sum_{j=1}^{\ell} (-1)^{j+1}\binom{m}{j} x_{\ell-j} t^{m-j}$ *for all* m,ℓ.

Proof Evaluate both sides at t^n, and the lemma reduces to checking the identity

$$\binom{m+n}{\ell} - \binom{n}{\ell} = \sum_{j=1}^{\ell} (-1)^{j+1}\binom{m}{j}\binom{m+n-j}{\ell-j} \text{ for all } m,n,\ell.$$

This is standard. □

PROPOSITION 2.9 D *is a simple ring.*

Proof Let $0 \neq I$ be a two-sided ideal of D. For some n, $I \cap D_n \neq 0$. A non-zero two sided ideal of a matrix ring over a ring R contains a non-zero ideal of R. Hence, for some n, $I \cap k[t^{p^n}] \neq 0$.

Choose $0 \neq f \in I \cap k[t]$, of lowest degree in t. Write $f = \alpha + g$ with $g \in tk[t]$, $\alpha \in k$. If $g = 0$ then $I \cap k \neq 0$, hence $I = D$, and the proof is complete. Suppose then, that $g \neq 0$, and let t^r be the lowest degree term appearing in g. Pick n, with $p^n \leq r < p^{n+1}$. Consider $[x_{p^n},f] = [x_{p^n},g] \in I$.

If $m \geq p^n$, then by Lemma 2.8, $[x_{p^n},t^m] = \sum_{j=1}^{p^n} (-1)^{j+1}\binom{m}{j}x_{p^n-j}t^{m-j} = (-1)^{p^n+1}\binom{m}{p^n}t^{m-p^n}$ since $\binom{m}{j} = 0 \pmod p$ for $j < p^n \leq m$. Also notice that as $p^n \leq r < p^{n+1}$ $\binom{r}{p^n} \neq 0 \pmod p$. So in particular $[x_{p^n},t^r] \neq 0$; thus $[x_{p^n},g]$ is of lower degree than f and is non-zero. This contradicts the choice of f. Thus $g = 0$, and the proof is complete. □

PROPOSITION 2.10 D *contains a proper subalgebra isomorphic to* D, *namely* $k[t^p,x_p,x_{2p},x_{3p},\ldots]$

Proof Notice that for all i,j $x_{ip}(t^{jp}) = \binom{jp}{ip}t^{(j-i)p}$ and that $\binom{jp}{ip} = \binom{j}{i} \pmod p$. Hence the natural action of x_{ip} on $k[t]$ maps $k[t^p]$ into $k[t^p]$, and so each x_{ip} is a differential operator on $k[t^p]$. After Theorem 2.6 $D(k[t^p]) = k[t^p,y_1,y_2,\ldots]$ where $y_i(t^{jp}) = \binom{j}{i}(t^p)^{j-i}$. As each x_{ip} acts as does y_i, we conclude that $D(k[t^p]) \cong k[t^p,x_p,x_{2p},\ldots]$; of course $D(k[t]) \cong D(k[t^p])$ so we have shown that $D \cong k[t^p,x_p,x_{2p},\ldots]$.

That $k[t^p,x_p,x_{2p},\ldots]$ is a proper subalgebra of D is obvious from the fact that $D = k[t] \oplus k[t]x_1 \oplus k[t]x_2 \oplus \ldots$ (this follows from Theorem 2.7) and $k[t^p,x_p,x_{2p},\ldots] = k[t^p] \oplus k[t^p]x_p \oplus \ldots$. □

The next example illustrates that one useful technique for studying the Weyl algebra in characteristic zero, is not available in characteristic p. If k is a field with char k = 0, then $D(A_k^1) \cong k[x,y]$ with $xy - yx = 1$; $D(A_k^1)$ can be localised at the non-zero elements of k[x] and k[y] respectively. The diagonal embedding of $D(\mathbb{A}_k^1)$ into the direct sum of the localisations, $D(\mathbb{A}_k^1) \to k(x)[y] \oplus k(y)[x]$, is a faithfully flat embedding; the "faithfulness" comes from the fact that if I is a left ideal of $D(\mathbb{A}_k^1)$ with $I \cap k[x] \neq 0$ and $I \cap k[y] \neq 0$ then, in fact, $I = D(\mathbb{A}_k^1)$.

<u>EXAMPLE</u> There is a left ideal I of D, $I \neq D$ such that $I \cap k[t] \neq 0$ and $I \cap k[x_1,x_2,\ldots] \neq 0$.

We construct our example for char k = 2, but a similar example exists for any characteristic.

So, assume p = 2, put $I = Dt^2 + Dx_1$. Recall, from Theorem 2.7 that D is a free left k[t]-module with basis $1,x_1,x_2,\ldots$, so $D = \bigoplus_{n=0}^{\infty} k[t]x_n$. Now $x_nx_1 = \binom{n+1}{1}x_{n+1} = \begin{cases} 0 & n \text{ odd} \\ x_{n+1} & n \text{ even} \end{cases}$, and thus $Dx_1 = \bigoplus_{n \text{ odd}} k[t]x_n$. As p = 2, $[t^2,x_1] = 0$, thus $(k[t] + k[t]x_1)t^2 \subseteq k[t]t^2 + k[t]x_1$. If $n \geq 2$, then $x_nt^2 = t^2x_n + x_{n-2}$, so $k[t]x_nt^2 = k[t](t^2x_n + x_{n-2})$. Consequently,

$$I \subseteq \sum_{n \text{ odd}} k[t]x_n + k[t]t^2 + \sum_{\substack{n \text{ even} \\ n \geq 2}} k[t](t^2x_n + x_{n-2}) =$$

$$k[t]t^2 + k[t]x_1 + k[t](t^2x_2+1) + k[t]x_3 + k[t](t^2x_4+x_2) + \ldots$$

and it is easy to see that $1 \notin I$.

<u>PROPOSITION 2.11</u> k[t] *is a simple* D-*module*.

<u>Proof</u> Let $0 \neq N$ be a submodule of k[t]. We will show $N \cap k \neq 0$ from which the result follows. Suppose $N \cap k = 0$, and choose $f \in N$ of least degree. Let t^r be the highest degree term appearing in f. Choose n such that $p^n \leq r < p^{n+1}$. Then then $x_{p^n}(t^r) = \binom{r}{p^n}t^{r-p^n}$, and $\binom{r}{p^n} \neq 0 \pmod p$. Hence $x_{p^n}(f) \neq 0$ and is of lower degree than f. This contradicts the choice of f. □

Recall that if k is of characteristic zero then the natural action of $k[t,\partial/\partial t]$ on $k[t]$ extends to an action of $k[t,\partial/\partial t]$ on $k[t,f^{-1}]$ for any $0 \neq f \in k[t]$, and that $k[t,f^{-1}]$ is of finite length as a $k[t,\partial/\partial t]$-module. The usual proof of this [2] uses Gelfand-Kirillov dimension. Although the same tool is no longer available in characteristic $p > 0$, the same result is true (Theorem 2.13). In order to prove this a few preliminary observations are required.

As $D_n \cong M_{p^n}(k[t^{p^n}])$, any non-zero D_n-module has dimension (over k) at least p^n. After Theorem 2.6 (and its proof) we have $D_n = k[t] \oplus x_1 k[t] \oplus \ldots \oplus x_{p^n-1} k[t]$ If $0 \neq f \in k[t]$ with $\deg(f) = F$ then $D_n/D_n f \cong S \oplus x_1 S \oplus \ldots \oplus x_{p^n-1} S$ where, $S = k[t]/(f)$, as a right $k[t]$-module. As $\dim_k S = F$, $\dim(D_n/D_n f) = p^n F$, and hence by our first observation $\text{length}_{D_n}(D_n f) \leq F$.

<u>LEMMA 2.12</u> *Let* M *be a left* D*-module, with a chain of finite dimensional subspaces* $M_0 \subset M_1 \subset M_2 \subset \ldots$ *such that*

(a) *each* M_n *is a* D_n*-module,*

(b) *for large* n, $\text{length}_{D_n}(M_n) \leq F$ (*fixed* F *for all* $n \gg 0$),

(c) $M = \bigcup_{n=0}^{\infty} M_n$.

Then, as a D*-module,* $\text{length}_D(M) \leq F$.

<u>Proof</u> Suppose F = 1. We must show that M is a simple D-module. Choose $0 \neq m \in M$ and choose any $m' \in M$. For all sufficiently large n, m and m' belong to M_n, which is a simple D_n-module by (b). Thus $m' \in D_n m \subset Dm$. Thus M is a simple D-module.

We now prove the result by induction on F. Suppose $F \geq 2$, and that the lemma is true for all numbers less than F. If M is simple as a D-module the proof is finished. If not, choose $0 \neq N$ a proper D-submodule of M. Put $N_n = N \cap M_n$; notice that $N = \bigcup_{n=0}^{\infty} N_n$, and each N_n is a D_n-module. We show that for all large n, $\text{length}_{D_n}(N_n) \leq F-1$. To see this, pick $m \in M$, $m \notin N$. There exists n_0 such that $m \in M_n$ for all $n \geq n_0$, but $m \notin N_n$. Hence, if $n \geq n_0$, $N_n \subsetneq M_n$. Thus $\text{length}_{D_n}(N_n) \leq F-1$ for all large n. By the induction hypotheses $\text{length}_D(N) \leq F-1$.

We have shown that any proper submodule of M has length at most F-1. Hence,

$\text{length}_D(M) \le F$. □

THEOREM 2.13 *Let* $0 \ne f \in k[t]$. *Then the* D*-module* $k[t,f^{-1}]$ *is of finite length (in fact, of length* $\le \deg(f) + 1$).

Proof   As $k[t]$ is a simple D-submodule of $k[t,f^{-1}]$, it is enough to show that $M = k[t,f^{-1}]/k[t]$ is of length $\le \deg(f)$.

For each n, let M_n be the D_n-submodule of M generated by the image of f^{-p^n}. If $gf^{-m} \in M$ with $g \in k[t]$, there exists an n, with $m < p^n$; then $gf^{-m} = gf^{p^n-m}f^{-p^n} \in M_n$. Hence $M = \bigcup_{n=0}^{\infty} M_n$.

Put $F = \deg(f)$. We will show that $\text{length}_{D_n}(M_n) \le F$, and the theorem will follow from Lemma 2.12. Recall that a non-zero D_n-module has dimension at least p^n, so it will suffice to show that $\dim_k M_n \le Fp^n$.

Recall that $D_n = k[t] \oplus k[t]x_1 \oplus \ldots \oplus k[t]x_{p^n-1}$, so if one has $x_j(f^{-p^n}) = 0$ for $1 \le j < p^n$, then $M_n = D_n.f^{-p^n} = k[t].f^{-p^n}$, and as $f^{p^n}.f^{-p^n} = 0$ (remember $M = k[t,f^{-1}]/k[t]$), it would follow that $\dim_k(M_n) = \dim_k(k[t]/\langle f^{p^n}\rangle) = Fp^n$.

So the theorem is complete if $x_j(f^{-p^n}) = 0$ for $1 \le j < p^n$. However, $f^{p^n} \in k[t^{p^n}]$, and as $x_j \in D_n$, x_j commutes with multiplication by f^{p^n}. Thus $x_j(f^{-p^n}) = f^{-p^n}x_j(1) = 0$, for $1 \le j < p^n$. □

The following is well known and is useful in deciding whether x_ix_j is zero or not.

LEMMA 2.14 *If* $a,b \in \mathbb{N}$ *and the p-adic expansions are* $a = a_o + a_1p + a_2p^2 + \ldots$, $b = b_o + b_1p + b_2p^2 + \ldots$ *then* $\binom{a}{b} \equiv \prod_{j=1} \binom{a_j}{b_j} \pmod p$.

LEMMA 2.15 *For* $m \ge n$, D_m *is free as a* D_n*-module (on either the right or the left) of rank* p^{m-n}. *A basis for* D_m *as a* D_n *-module is given by* $1, x_{p^n}, x_{2p^n}, \ldots, x_{(p^m-1)p^n}$.

Proof   Recall the description of D_n and D_m given in Theorem 2.6. If $0 \le j \le p^n-1$, and $0 \le i \le p^m-1$ then $x_jx_{ip^n} = \binom{j+ip^n}{j}x_{j+ip^n}$. However, writing j and ip^m in their p-adic form, Lemma 2.14 ensures that $x_jx_{ip^n} \ne 0$. The Lemma follows. □

The following consequence of Lemma 2.12 is useful.

<u>LEMMA 2.16</u> *If* N *is a* D_n*-module of finite length, then* $D \otimes_{D_n} N$ *is of finite length as a* D*-module.*

<u>Proof</u> If N were a faithful D_n-module then D_n would be artinian (which it is not). So $I = \mathrm{ann}_{D_n}(N) \neq 0$. But a non-zero ideal of $D_n = M_{p^n}(k[t^{p^n}])$ intersects $k[t^{p^n}]$ in a non-zero ideal. Thus N is a finitely generated module over the finite dimensional algebra $M_{p^n}(k[t^{p^n}]/I \cap k[t^{p^n}]) = D_n/I$. Thus $\dim_k N < \infty$.

Let $m \geq n$. As D_m is a free D_n-module of rank p^{m-n}, $D_m \otimes_{D_n} N$ is of dimension $\leq p^{m-n} \dim_k N$. As a non-zero D_m-module has dimension $\geq p^m$, $\mathrm{length}_{D_m}(D_m \otimes_{D_n} N) \leq p^{-n} \dim_k N$. The lemma follows from Lemma 2.12 by observing that $D \otimes_{D_n} N = \bigcup_{m \geq n} D_m \otimes_{D_n} N$. □

We next show that gl.dim.D = 1. As the comments and example following Proposition 2.10 indicate, the proof that $\mathrm{gl.dim.}(D(A_k^1)) = 1$ when k is of characteristic zero cannot be used. The following preparatory lemma is required (and allows us in the proof of Theorem 2.18 to make frequent use of the fact that for a finitely generated D_n-module the concepts of torsion submodule coincide whether we consider torsion with respect to the regular elements of D_n, or with respect to the non-zero elements of k[t] when the D_n-module is viewed as a k[t]-module).

<u>LEMMA 2.17</u> *Let* M *be a finitely generated* D_n*-module. Let* M_1 *be the torsion submodule of* M *with respect to the regular elements of* D_n*; let* M_2 *be the torsion submodule of* M *with respect to* $k[t^{p^n}]$*; let* M_3 *be the torsion submodule of* M *with respect to* k[t]. *Then* $M_1 = M_2 = M_3$.

<u>Proof</u> As $k[t] \subset D_n$ and D_n is a free k[t]-module, $k[t]\backslash\{0\}$ consists of regular elements in D_n. Hence $M_3 \subset M_1$. Similarly $M_2 \subset M_3 \subset M_1$.

Write Q_n for the ring of fractions of D_n. That is, $Q_n = M_{p^n}(k(t^{p^n})) = k(t^{p^n}) \otimes_{k[t^{p^n}]} D_n$, where $k(t^{p^n})$ denotes the field of rational functions in t^{p^n}. Now $Q_n \otimes_{D_n} M_1 = 0$. Hence $k(t^{p^n}) \otimes_{k[t^{p^n}]} M_1 = 0$, and it follows that $M_1 \subset M_2$. □

<u>THEOREM 2.18</u> gl.dim. D = 1.

Proof As D is not semi-simple artinian, gl.dim. $D \geq 1$. So it is enough to show that every left ideal of D is projective. Let I be a left ideal.

Put $I_n = I \cap D_n$, and define I'_n to be the left ideal of D_n containing I_n such that I'_n/I_n is the torsion submodule of the D_n-module D_n/I_n. Put $T_n = DI'_n \cap I$.

We claim that $T_n \subset T_{n+1}$. To see this it is enough to check that $I'_n \subset I'_{n+1}$. But $I'_n + I_{n+1}/I_{n+1} \cong I'_n/I'_n \cap I_{n+1}$ which is a homomorphic image of I'_n/I_n. As I'_n/I_n is k[t]-torsion so is $I'_n + I_{n+1}/I_{n+1}$. Thus $I'_n \subset I'_{n+1}$.

We claim that T_n is a finitely generated left ideal. Notice that $T_n/DI_n \subseteq DI'_n/DI_n \cong D \otimes_{D_n} (I'_n/I_n)$. By Lemma 2.16 this latter D-module is of finite length since I'_n/I_n is of finite length as a D_n-module. The truth of the claim follows from the fact that DI_n is finitely generated, and that T_n/DI_n is of finite length.

Consider T_{n+1}/T_n. As both these left ideals are finitely generated there exists $m \in \mathbb{N}$ with $T_{n+1}/T_n = D(T_{n+1} \cap D_m)/D(T_n \cap D_m)$. Now $T_{n+1} \cap D_m/T_n \cap D_m \cong T_n + (T_{n+1} \cap D_m)/T_n$ which is a submodule of $I/T_n = I/I \cap DI'_n \cong I + DI'_n/DI'_n$ which is a submodule of $D/DI'_n \cong D \otimes_{D_n} (D_n/I'_n)$. However, as a k[t]-module D_n/I'_n is torsion-free, hence so is D/DI'_n. Thus $T_{n+1} \cap D_m/T_n \cap D_m$ is torsion-free as a D_m-module. But D_m is a hereditary Noetherian prime ring, so by [5, Theorem 2.1] a torsion-free D_m-module is projective. Hence there is a left ideal J of D_m with $T_{n+1} \cap D_m = T_n \cap D_m \oplus J$. Thus (as D is free as a D_m-module) $D(T_{n+1} \cap D_m) = D(T_n \cap D_m) \oplus DJ$. In particular, there is a finitely generated left ideal S_n with $T_{n+1} = T_n \oplus S_n$.

Now $I = \bigcup_{n=0}^{\infty} DI_n = \bigcup_{n=0}^{\infty} T_n = T_0 + T_1 + \ldots = S_0 \oplus S_1 \oplus S_2 \oplus \ldots$. But each S_n is finitely generated hence projective (because $S_n \cong D \otimes_{D_m} (D_m \cap S_n)$, and $D_m \cap S_n$ is a projective D_m-module). Thus I is projective. □

Goodearl has pointed out the following way of viewing D. Let B denote the subring $k[x_1, x_2, \ldots]$ of D; B is isomorphic to the factor ring of a commutative polynomial ring $k[X_1, X_2, \ldots]$ modulo the ideal generated by $X_iX_j - \binom{i+j}{i}X_{i+j}$. The inner derivation $ad(t) = [t,-]$ of D maps B into itself, so ad(t) acts as a derivation on B, and D may be viewed as B[t], the extension of B by the derivation ad(t). Now it is easy to see gl.dim. $B = \infty$ because there exist non-split exact sequences:

$$0 \to x_1 B \to B \to x_{p-1}B \to 0$$

$$0 \to x_{p-1}B \to B \to x_1 B \to 0.$$

Hence this gives an example of a commutative ring of infinite global dimension such that an extension by a derivation has finite global dimension (the first such example appears in [6]).

As D is a ring of differential operators it has a filtration given by the order of the operators. As x_n is of order n, the filtration is given by $F_n D = k[t] \oplus k[t]x_1 \oplus \ldots \oplus k[t]x_n$, and the associated graded algebra grD is isomorphic to B[s] where s is a commuting indeterminate. Hence although gl.dim D = 1, gl.dim (grD) = ∞.

Notice that the exact sequences over D corresponding to those for B given above are split. This is because Dx_{p-1} is projective (being generated by the idempotent $t^{p-1}x_{p-1}$).

We now briefly turn our attention to the ring of fractions of D. As D is a free k[t]-module, k[t]\{0} consists of regular elements of D. Hence Fract D contains k(t). As $D_n \cong M_{p^n}(k[t^{p^n}])$, Fract $D_n \cong M_{p^n}(k(t^{p^n}))$. Thus we have

THEOREM 2.19 *The ring of fractions* Q, *of* D, *is equal to* $k(t)[x_1,x_2,\ldots]$ *and* $Q = \bigcup_{n=0}^{\infty} Q_n$ where $Q_n = \mathrm{End}_{k(t^{p^n})}k(t) = \mathrm{Fract}\ D_n$.

In particular Q is a union of simple artinian rings, so is von Neumann regular. As Q is flat as a D-module, gl.dim. Q $\leq$ gl.dim. D. But Q is not semi-simple artinian, so gl. dim. Q = 1.

PROPOSITION 2.20 Q *is not self-injective.*

Proof It is sufficient to find a left ideal J of Q, and a Q-module map $\phi: J \to Q$ which is not the restriction of a Q-module map $\psi: Q \to Q$.

Put $J = Qx_1 + Qx_2 + \ldots$; consider the formal sum $y = \sum_{j=0}^{\infty} x_{p^j-1}$, and define $\phi: J \to Q$ by $\phi(r) = ry$. This does make sense: notice that $x_i x_{p^j-1} = 0$ if i is fixed and j is sufficiently large, thus ry is actually a finite sum for $r \in J$. So ϕ is a bona-fide Q-module homomorphism.

Suppose that $\psi: Q \to Q$ is a left Q-module map. Then ψ is just right multiplication by $z = \psi(1)$. So if $\phi = \psi|_J$ then, in particular, $x_i(y-z) = 0$ for all $i \geq 1$. Suppose $z = a_0 + x_1 a_1 + \ldots + x_n a_n$ with each $a_j \in k[t]$, and $a_n \neq 0$. Suppose $p^m-1 > n$. Then $x_{p^m}.y = \sum_{j=0}^{m} x_{p^m}.x_{p^j-1}$, and $x_{p^m}.x_{p^m-1} \neq 0$, but $x_{p^m}z$ cannot contain a term involving x_{2p^m-1} since $n < p^m-1$. Hence $x_{p^m}.y \neq x_{p^m}z$, and thus $\phi \neq \psi|_J$. □

§3. PROPERTIES OF $D(\mathbb{P}_p^1)$

We begin by defining $D(\mathbb{P}_p^1)$. Let $\mathcal{D}$ be the sheaf of differential operators on $\mathbb{P}^1$, and define $D(\mathbb{P}^1) = \Gamma(\mathbb{P}^1, \mathcal{D})$. As $\mathcal{D}$ is the unique quasi-coherent sheaf of $\mathcal{O}_{\mathbb{P}^1}$-modules such that for every open affine $U \subset \mathbb{P}^1$, $\Gamma(U,\mathcal{D})$ is the ring of differential operators on $\mathcal{O}(U)$ (the ring of regular functions on U) to compute the global sections of $\mathcal{D}$ we may proceed as follows. Let U_+, U_- be two copies of $\mathbb{A}^1$ covering $\mathbb{P}^1$ such that $\mathcal{O}(U_+) = k[t]$, $\mathcal{O}(U_-) = k[t^{-1}]$ and let D^+, D^- denote the rings of differential operators on U^+ and U^- respectively. If D^+ and D^- are considered as subalgebras of $D(k(t))$, we have $D(\mathbb{P}^1) = D^+ \cap D^-$. As $D^+ = \{\theta \in D(k(t)) | \theta(k[t]) \subset k[t]\}$ and $D^- = \{\theta \in D(k(t)) | \theta(k[t^{-1}]) \subset k[t^{-1}]\}$ we have $D(\mathbb{P}^1) = \{\theta \in D(k(t)) | \theta(k[t]) \subset k[t]$ and $\theta(k[t^{-1}]) \subset k[t^{-1}]\}$. Thus we obtain (for k a field of characteristic $p > 0$)

LEMMA 3.1 *Fix* n, *put* $q = p^n$ *and let* $\theta \in D_n^+$ (using the notation of §2). *Then* $\theta \in D(\mathbb{P}_k^1)$, *if and only if*

(1) $\theta(1) \in k$

(2) $\theta(t^j) \in$ lin.span $\langle 1, t, t^2, \ldots, t^q \rangle$ *for all* j, $0 < j < q$.

Proof Suppose θ satisfies the conditions. First observe that θ extends to a $k[t^q]$-linear differential operator on $k(t)$ (since $\theta \in D_n^+$). Pick $i > 0$; we show that $\theta(t^{-i}) \in k[t^{-1}]$.

Pick m such that $mq < i \le (m+1)q$. Then $0 \le (m+1)q-i < q$, so by (2), $\theta(t^{(m+1)q-i}) \in$ lin.span $<1,t,t^2,\dots,t^q>$. But $\theta(t^{(m+1)q-i}) = t^{(m+1)q}\theta(t^{-i})$, hence $\theta(t^{-i}) \in$ lin.span $t^{-(m+1)q}<1,t,\dots,t^q> \subset k[t^{-1}]$. This and (1) ensure that $\theta(k[t^{-1}]) \subset k[t^{-1}]$, and so $\theta \in D(\mathbb{P}^1_k)$. The conditions are therefore sufficient.

On the other hand, if $\theta \in D(\mathbb{P}^1_k)$, then certainly $\theta(k[t] \cap k[t^{-1}]) \subset k[t] \cap k[t^{-1}]$, so (1) is necessary. Also if $0 < j < q$, then $\theta(t^{-j}) \in k[t^{-1}]$, and hence $\theta(t^{q-j}) = t^q\theta(t^{-j}) \in t^q k[t^{-1}] \cap k[t] =$ lin.span. $<1,t,\dots,t^q>$. So (2) is necessary. □

Put $D(\mathbb{P}^1)_n = D(\mathbb{P}^1) \cap D_n^+$; that is, $D(\mathbb{P}^1)_n$ is the differential operators in $D(\mathbb{P}^1)$ of order $\le n$. Notice that after the lemma, $\dim_k D(\mathbb{P}^1)_n = 1 + (p^n-1)(p^n+1) = p^{2n}$, so $D(\mathbb{P}^1)$ is a union of finite dimensional subalgebras.

<u>LEMMA 3.2</u> *The nilpotent radical of* $D(\mathbb{P}^1)_n$ *is the span of those* θ *which satisfy*

(1) $\theta(1) = 0$

(2) $\theta(t^j) \in$ lin.span $<1,t^{p^n}>$ *for all* $0 < j < p^n$.

<u>Proof</u>

Put $q = p^n$. First the span of such θ is an ideal of $D(\mathbb{P}^1)_n$. If $\psi \in D(\mathbb{P}^1)_n$, then $\psi\theta(1) = \theta\psi(1) = 0$; and for $0 < j < q$, one has $\psi\theta(t^j) \subset$ lin.span $<\psi(1),\psi(t^q)> =$ lin.span $<\psi(1), t^q\psi(1)> \subset$ lin.span $<1,t^q>$ by Lemma 3.1(1); also $\theta\psi(t^j) \subset$ lin.span $<\theta(1),\theta(t),\dots,\theta(t^q)> \subset$ lin.span $<1,t^q>$ as $\theta(t^q) = t^q\theta(1) = 0$. We have shown that if θ satisfies (1) and (2), so do $\theta\psi$ and $\psi\theta$. Hence the span of such θ is an ideal.

The square of this ideal is zero: if θ and ψ satisfy (1) and (2) then $\psi\theta(1) = 0$ and for $0 < j < q$, $\psi\theta(t^j) \subset$ lin.span $<\psi(1),\psi(t^q)> = 0$.

The factor by this ideal is semi-simple artinian: the factor may be identified with those θ such that $\theta(1) \in k$ and $\theta(t^j) \in$ lin.span $<t,t^2,\dots,t^{q-1}>$ for $1 \le j < q$; but this algebra is isomorphic to $(\mathrm{End}_k k) \oplus (\mathrm{End}_k k^{q-1})$. □

Put N_n = nilpotent radical of $D(\mathbb{P}^1)_n$; notice that $\dim N_n = 2(p^n-1)$.

<u>LEMMA 3.3</u> $N_n \cap N_{n+1} = 0$.

<u>Proof</u> Pick $0 \ne \theta \in N_n$. Then $\theta(t^j) \ne 0$ for some $0 < j < p^n$. Hence, if $\theta \in N_{n+1}$,

then $\theta(t^j) \in$ lin. span $<1,t^{p^{n+1}}> \cap$ lin.span $<1,t^{p^n}> = k$. But $0 < j + p^n < p^{n+1}$ and $\theta(t^{j+p^n}) = t^{p^n}\theta(t^j) \in kt^{p^n}$. But by applying Lemma 3.2(2) for n+1, one must have $\theta(t^{j+p^n}) \in$ lin.span $<1,t^{p^{n+1}}>$. Thus $\theta(t^{j+p^n}) = 0$, whence $\theta(t^j) = 0$. This contradiction gives the result. □

PROPOSITION 3.4 $D(\mathbb{P}^1)$ *contains no non-zero nilpotent ideal.*

Proof Suppose $N \neq 0$, is a nilpotent ideal. Then $N \cap D(\mathbb{P}^1)_n \neq 0$ for some n. Thus $N \cap D(\mathbb{P}^1)_n$ is a nilpotent ideal of D_n. Similarly $N \cap D(\mathbb{P}^1)_{n+1}$ is a nilpotent ideal of $D(\mathbb{P}^1)_{n+1}$. Hence $0 \neq N \cap D(\mathbb{P}^1)_n \subset N_n \cap N_{n+1}$. This contradicts Lemma 3.3. □

PROPOSITION 3.5 $D(\mathbb{P}^1)$ *is not von Neumann regular.*

Proof Consider $x_1 \in D^+$ (the notation is that of §2). One sees that $x_1 = \partial/\partial t \in D(\mathbb{P}^1)$. Suppose there exists $a \in D(\mathbb{P}^1)$ with $x_1 a x_1 = x_1$. Then in particular, as $x_1(t) = 1$, one has $x_1 a(1) = 1$. But if $a \in D(\mathbb{P}^1)$ then $a(1) = 1$. However, $x_1(k) = 0$, so there exists no $a \in D(\mathbb{P}^1)$ with $x_1 a(1) = 1$. Hence the result. □

PROPOSITION 3.6 $D(\mathbb{P}^1)$ *is its own ring of fractions.*

Proof This is true of any algebra which is a union of finite dimensional algebras over a field (since an artinian ring is its own ring of fractions). □

PROPOSITION 3.7 (1) $D(\mathbb{P}^1)_n$ *is the sum of the two-sided ideals* $J_n = \{\theta \in D(\mathbb{P}^1)_n \mid \theta(t^j) \in k$ for all $0 \le j < p^n\}$ *and* $Q_n = \{\theta \in D(\mathbb{P}^1)_n \mid \theta(1) = 0\}$. (2) $\dim_k (D(\mathbb{P}^1)_n/Q_n) = 1$ (3) $J_n \cap Q_n = N_n$. (4) For $n \ge 1$, J_n/N_n *and* Q_n/N_n *are minimal ideals of* $D(\mathbb{P}^1)_n/N_n$. (5) *Let* $\alpha \in D(\mathbb{P}^1)_n$. *The two sided ideal of* $D(\mathbb{P}^1)_n$ *generated by* α *equals* $D(\mathbb{P}^1)_n$ *if and only if* α *can be written in the form* $\alpha = \beta + \gamma$ *with* $\beta \in J_n \setminus N_n$ *and* $\gamma \in Q_n \setminus N_n$.

Proof After Lemmas 3.1 and 3.2 the proposition is straightforward. □

PROPOSITION 3.8 (Notation as in (3.7)). *Put* $Q = \bigcup_{n=0}^{\infty} Q_n$. *Then Q is the unique proper ideal of* $D(\mathbb{P}^1)$, *and* $D(\mathbb{P}^1)/Q \cong k$.

Proof As each $Q_n \subset Q_{n+1}$, and Q_n is an ideal of $D(\mathbb{P}^1)_n$, Q is a two sided ideal of

$D(\mathbb{P}^1)$.

Suppose $\theta \in D(\mathbb{P}^1)_n$ and $\theta \notin Q_n$. Then $D(\mathbb{P}^1)\theta D(\mathbb{P}^1) = D(\mathbb{P}^1)$. To prove this it is enough to show that $D(\mathbb{P}^1)_{n+1}\theta D(\mathbb{P}^1)_{n+1} = D(\mathbb{P}^1)_{n+1}$. As $\theta \notin Q_n$, $\theta(1) \neq 0$. Hence, without loss of generality $\theta(1) = 1$. As θ is $k[t^{p^n}]$-linear, $\theta(t^{p^n}) = t^{p^n}$, and it follows that $\theta \notin J_{n+1}$, and $\theta \notin Q_{n+1}$. Hence by Proposition 3.7(5), the two sided ideal of $D(\mathbb{P}^1)_{n+1}$ generated by θ is $D(\mathbb{P}^1)_{n+1}$ itself.

It follows that any two sided ideal of $D(\mathbb{P}^1)$ not equal to $D(\mathbb{P}^1)$ must be contained in Q.

Suppose now that $\theta \in Q$, $\theta \neq 0$. We show θ generates Q. Suppose $\theta \in D(\mathbb{P}^1)_n$. Hence $\theta(1) = 0$, and as $\theta \neq 0$, $\theta(t^j) \neq 0$ for some j, $0 < j < p^n$. Hence $\theta(t^{j+p^n}) = t^{p^n}\theta(t^j) \notin k$. Thus $\theta \notin J_{n+1}$. It follows that $D(\mathbb{P}')_{n+1}\theta D(\mathbb{P}')_{n+1} = Q_{n+1}$. This is true for all $n \gg 0$, so $D(\mathbb{P}^1)\theta D(\mathbb{P}^1) = Q$.

Thus Q is the unique proper ideal of $D(\mathbb{P}^1)$. Finally as $\dim_k(D(\mathbb{P}^1)_n/Q_n) = 1$ for all n, $\dim_k(D(\mathbb{P}^1)/Q) = 1$. □

PROPOSITION 3.9 $D(\mathbb{P}^1)$ *is a primitive ring, and* $k[t]$ *is a faithful module of length 2, the submodule being* k.

Proof This is an immediate consequence of Lemma 3.1. □

We now compute $K_0(D(\mathbb{P}^1))$. As K_0 commutes with direct limits, one has $K_0(D(\mathbb{P}^1)) = \varinjlim K_0(D(\mathbb{P}^1)_n)$. We need only consider $n \geq 1$, so henceforth assume $n \geq 1$.

Recall that $D(\mathbb{P}^1)_n/N_n = J_n/N_n \oplus Q_n/N_n$ and $J_n/N_n \cong k$ while $Q_n/N_n \cong M_{p^n-1}(k^{p^n-1})$ (this is implicit in the proof of Lemma 3.2). Hence $K_0(D(\mathbb{P}^1)_n) = \mathbb{Z} \oplus \mathbb{Z}$ with $[D(\mathbb{P}^1)_n] = (1,p^n-1)$. The positive cone in $K_0(D(\mathbb{P}^1)_n)$ is $K_0^+(D(\mathbb{P}^1)_n) = \{(a,b) \in \mathbb{Z} \oplus \mathbb{Z} \mid a \geq 0,\ b \geq 0\}$.

The embedding $D(\mathbb{P}^1)_n \to D(\mathbb{P}^1)_{n+1}$ induces maps $\phi_n : K_0(D(\mathbb{P}^1)_n) \to K_0(D(\mathbb{P}^1)_{n+1})$ given by $\phi_n(1,0) = (1,p-1)$ and $\phi_n(0,1) = (0,p)$.

Define $G_n = \mathbb{Z} \oplus \mathbb{Z}$ and let $\psi_n : G_n \to G_{n+1}$ be the group homomorphism $\psi_n(1,0) = (1,0)$, $\psi_n(0,1) = (0,p)$. Define $\delta : \mathbb{Z} \oplus \mathbb{Z} \to \mathbb{Z} \oplus \mathbb{Z}$ by $\delta(1,0) = (1,1)$, $\delta(0,1) = (0,1)$, and extend δ to a group isomorphism. Then $\delta : (K_0(D(\mathbb{P}^1)_n, \phi_n) \to (G_n, \psi_n)$ is a chain isomorphism, so $K_0(D(\mathbb{P}^1)) = \varinjlim (G_n,\psi_n)$. As ψ_n is just the multiplication map $(a,b) \xrightarrow{(1,p)} (a,bp)$ one sees that this direct limit is $\mathbb{Z} \oplus \mathbb{Z}[1/p]$, and that $[D(\mathbb{P}^1)] = (1,p)$.

By chasing the positive cones $K_0^+(D(\mathbb{P}^1)_n)$, one obtains $K_0^+(D(\mathbb{P}^1)) = \{(a,b) \in \mathbb{Z} \oplus \mathbb{Z}[1/p] \mid a \geq 0 \text{ and } b > 0 \text{ or } (a,b) = (0,0)\}$. It is an easy matter now to see that the only order ideal in $K_0(D(\mathbb{P}^1))$ apart from 0 and $K_0(D(\mathbb{P}^1))$ is $\mathbb{Z}[1/p]$.

Hence the lattice of order ideals is isomorphic to the lattice of two sided ideals of $D(\mathbb{P}^1)$. We summarise the above.

<u>THEOREM 3.10</u> $K_0(D(\mathbb{P}^1)) \cong \mathbb{Z} \oplus \mathbb{Z}[1/p]$, *with* $[D(\mathbb{P}^1)] = (1,p)$. *The lattice of order ideals in* $K_0(D(\mathbb{P}^1))$ *is isomorphic to the lattice of two sided ideals in* $D(\mathbb{P}^1)$; *this lattice is:*

<u>Remark</u> In [7, Corollary 15.21] it is proved that if R is a unit-regular ring there is an isomorphism between the lattice of two sided ideals of R, and the order ideals of $K_0(R)$. Of course after Proposition 3.5, $D(\mathbb{P}^1)$ is not unit-regular.

Recall that if k is a field of characteristic zero, then there is a surjective map $U(sl(2,k)) \to D(\mathbb{P}^1_k)$. This map is given by $e \to t^2\partial/\partial t$, $f \to -\partial/\partial t$, $h \to 2t\partial/\partial t$ where $e = \begin{pmatrix}0&1\\0&0\end{pmatrix}$, $f = \begin{pmatrix}0&0\\1&0\end{pmatrix}$, $h = \begin{pmatrix}1&0\\0&-1\end{pmatrix}$ is the usual basis for $sl(2,k)$. The surjectivity is seen from the fact that $D(\mathbb{P}^1_k) = k[\partial/\partial t, t\partial/\partial t, t^2\partial/\partial t]$, and this equality can be proved by elementary arguments. We show below that, if char $k = p > 0$, then the analogous map does not give a surjection from U_k, the hyperalgebra of $sl(2,k)$, to $D(\mathbb{P}^1_k)$.

So k is once again a field of characteristic $p > 0$. Denote the $\mathbb{Z}$-span of the elements $\frac{f^a}{a!}\binom{h}{b}\frac{e^c}{c!}$ with $a,b,c \in \mathbb{N}$, in $U(sl(2,\mathbb{C}))$ by $U_{\mathbb{Z}}$; this is the Kostant $\mathbb{Z}$-form and is a $\mathbb{Z}$-subalgebra of $U(sl(2,\mathbb{C}))$. The hyperalgebra U_k is defined to be $U_k = k \otimes_{\mathbb{Z}} U_{\mathbb{Z}}$.

$D(\mathbb{P}^1_{\mathbb{Z}})$ is equal to $D(\mathbb{Z}[t]) \cap D(\mathbb{Z}[t^{-1}])$, the intersection being taken inside $D(\mathbb{Z}[t,t^{-1}])$. Hence $D(\mathbb{P}^1_{\mathbb{Z}})$ is precisely those elements of $D(\mathbb{P}^1_{\mathbb{C}})$ which, when acting on $\mathbb{C}[t]$ and $\mathbb{C}[t^{-1}]$, map $\mathbb{Z}[t]$ into $\mathbb{Z}[t]$ and $\mathbb{Z}[t^{-1}]$ into $\mathbb{Z}[t^{-1}]$. The image of $\frac{f^a}{a!}\binom{h}{b}\frac{e^c}{c!}$ in $D(\mathbb{P}^1_{\mathbb{C}})$ is of course $\frac{(-\partial/\partial t)^a}{a!}\binom{2t\,\partial/\partial t}{b}\frac{(t^2\partial/\partial t)^c}{c!}$, and it is easy to check that this differential operator sends $\mathbb{Z}[t]$ to $\mathbb{Z}[t]$ and $\mathbb{Z}[t^{-1}]$ to $\mathbb{Z}[t^{-1}]$. Hence this element belongs to $D(\mathbb{P}^1_{\mathbb{Z}})$. Thus the map $U(sl(2,\mathbb{C})) \to D(\mathbb{P}^1_{\mathbb{C}})$ restricts to give a map $U_{\mathbb{Z}} \to D(\mathbb{P}^1_{\mathbb{Z}})$. This in turn induces a map $\phi: U_k \to D(\mathbb{P}^1_k)$ since $D(\mathbb{P}^1_k) = k \otimes_{\mathbb{Z}} D(\mathbb{P}^1_{\mathbb{Z}})$. This last equality derives from Theorem 2.7.

THEOREM 3.11 *The map $\phi: U_k \to D(\mathbb{P}^1_k)$ is not surjective.*

Proof Give $k[t,t^{-1}]$ the grading where t is of degree 1; define $D(\mathbb{P}^1_k)(j) = \{\theta \in D(\mathbb{P}^1_k) \mid \theta(kt^i) \subset kt^{i+j}$ for all $i \in \mathbb{Z}\}$. Then $D(\mathbb{P}^1_k) = \bigoplus_{j\in\mathbb{Z}} D(\mathbb{P}^1_k)(j)$ and this gives a grading on $D(\mathbb{P}^1_k)$. Notice that $\phi(e) \in D(\mathbb{P}^1_k)(1)$, $\phi(f) \in D(\mathbb{P}^1_k)(-1)$, $\phi(h) \in D(\mathbb{P}^1_k)(0)$. Likewise, $\phi(\frac{f^a}{a!}\binom{h}{b}\frac{e^c}{c!}) \in D(\mathbb{P}^1_k)(c-a)$.

Consider the element $t^{p-1}\frac{(\partial/\partial t)^p}{p!}$ which belongs to $D(\mathbb{P}^1_k)(1)$. We will show this is not in the image of ϕ. If it were in the image of ϕ, then it would be a linear combination of the image of elements $\frac{f^a}{a!}\binom{h}{b}\frac{e^c}{c!}$ with $c-a = 1$. Notice that $t^{p-1}\frac{(\partial/\partial t)^p}{p!}$ acts on $k[t]$ sending t^p to t^{p-1}. The action of $\frac{(\partial/\partial t)^a}{a!}\binom{2t\partial/\partial t}{b}\frac{(t^2\partial/\partial t)^{a+1}}{(a+1)!}$ sends t^p to $\binom{p+a}{p-1}\binom{2p+2a+2}{b}\binom{p+a+1}{p-1}t^{p-1}$. However, for all $a \in \mathbb{N}$, $\binom{p+a}{p-1}\binom{p+a+1}{p-1} \equiv 0 \pmod p$. Hence $\phi(\frac{f^a}{a!}\binom{h}{b}\frac{e^{a+1}}{(a+1)!})$ sends t^p to zero. Consequently, no linear combination of these elements can equal $t^{p-1}\frac{(\partial/\partial t)^p}{p!}$ which sends t^p to t^{p-1}. $\square$

REFERENCES

[1] A. Beilinson and J.N. Bernstein, Localisation de g-modules, *C.R. Acad. Sci.* 292 (1981) 15-18.

[2] J.N. Bernstein, The analytic continuation of generalized functions with respect to a parameter, *Funkcional. Anal. i. Prilozen* 6 (1972) 26-40.

[3] J. Dixmier, Sur les algèbres de Weyl, *Bull. Soc. Math. Fr.* 96 (1968) 209-242.

[4] J. Dixmier, Sur les algèbres de Weyl II, *Bull. Sci. Math.* 94 (1970) 289-301.

[5] D. Eisenbud and J.C. Robson, Modules over Dedekind prime rings, *J. Algebra* 16 (1970) 67-85.

[6] K. Goodearl, Global dimension of differential operator rings, *Proc. A.M.S.* 45 (1974) 315-322.

[7] K. Goodearl, *Von Neumann Regular rings*, Pitman (1979).

[8] A. Grothendieck, Eléments de Geometrie Algèbrique IV, Inst. des Hautes Études Sci., Publ. Math. No. 32 (1967).

[9] R.G. Heynemann and M. Sweedler, Affine Hopf Algebras, *J. Algebra* 13 (1969) 192-241.

10] T. Levasseur, Anneaux d'opérateurs differentiels, Seminaire M.P. Malliavin, Lecture Notes in Mathematics, No. 867, Springer-Verlag (1980).

NONCOMMUTATIVE LOCALIZED RINGS

R.B. Warfield, Jr.

For a large class of noncommutative Noetherian rings, there is now a reasonable notion of the localization at a prime ideal. The localized rings, however, are quite unlike local rings, and may have a countably infinite family of maximal ideals. The main purpose of this paper is to show to what extent the localized rings are nonetheless like local rings, mostly in K-theoretic terms. We also give a proof of a localization theorem and show, as an application, that if R is a fully bounded Noetherian ring which is an algebra over an uncountable field, then gl. dim $(R) = sup$ {pr. dim. S : S a simple right R-module }. (It is unknown whether such a result holds for Noetherian rings in general.)

If P is a prime ideal in a Noetherian ring R, then there is a subset $Cl(P)$ of Spec (R), the *clique* associated to P, which is a finite or countable set and consists of primes "linked" to P in a way described below. For a certain class of Noetherian rings (including fully bounded Noetherian rings which are algebras over an uncountable field, and group rings $K(G)$ with G polycyclic-by-finite and K an uncountable field) it is possible to show that R has a localization $\overline{R}$ associated to this clique, such that the primitive ideals of $\overline{R}$ are all co-Artinian and are in one-to-one correspondence with the primes in $Cl(P)$. After preliminary results in section 1, we give a proof of the localization theorem in section 2. In section 3, we make a more careful study of the localized rings that arise and show that in many respects they behave like local or semilocal rings. One way of thinking about this is to say that they have "zero-dimensional" K-theory, even though their Krull and J-dimensions are usually greater than zero. For example, we show (Theorem 11) that if R is such a localized ring, then $K_0(R)$ is free Abelian, of at most countable rank.

It is not known whether for an arbitrary Noetherian ring R, gl. dim. R = *sup* { pr. dim. S : S a simple right R-module }. One can show this if one assumes in advance that the global dimension of R is finite ([3], [10, Thm. 16]), so the real question is whether the simple modules can be used to determine whether the global dimension is finite or infinite. We show (Theorem 12) that this holds for the localized rings arising from this theory, and, as an application, we show that the same result holds for any fully bounded Noetherian ring which is an algebra over an uncountable field (Corollary 14).

In this paper, all modules are right modules unless otherwise specified. Properties such as "Noetherian" and "Goldie" are intended to hold on both sides unless specified otherwise. A prime ring R is *right bounded* if every essential right ideal contains a nonzero two-sided ideal. A ring is *fully bounded Noetherian* (FBN) if for every prime ideal P of R, R/P is right and left bounded.

The author's research was supported in part by a grant from the National Science Foundation.

1. Elements regular at families of prime ideals.

If R is a Noetherian ring and $r \in R$, we say r is *regular at the prime* P if $r+P$ is a regular element (non-zero-divisor) in R/P. The set of elements regular at P is written $C(P)$. To consider regularity at different primes, it is convenient to put a topology on the set Spec (R) of prime ideals of R. There are two such topologies which are frequently used -- the *Zariski topology* and the *patch topology*. In the Zariski topology, the closed sets are the sets

$$V(I) = \{P \in \mathrm{Spec}(R):\ P \supseteq I\}$$

for all ideals I of R. Similarly, the open sets are the sets

$$W(I) = \{P \in \mathrm{Spec}(R):\ P \not\supseteq I\}\ .$$

The patch topology is obtained by taking as a sub-basis of closed sets all of the closed sets and all of the quasi-compact open sets in the Zariski topology [13]. If the ring R satisfies the ascending chain condition on semiprime ideals (e.g., a Noetherian ring or an affine P.I. ring), then it is easy to see that the sets $W(I)$ are all quasi-compact. It follows that in the patch topology, the sets $W(I)$ and $V(I)$ are all open, and a typical neighborhood of a point P is given by $V(P) \cap W(I)$ for some ideal I. With respect to this topology, Spec (R) becomes a compact Hausdorff space. A prime ideal is said to be a J-prime if it is the intersection of primitive ideals. The set of J-primes is written J-Spec (R), and it is easy to verify that it is a closed (and therefore, compact) subset of Spec (R).

Definition. If R is a Noetherian ring and X is a subset of Spec (R), then X satisfies the *generic regularity condition* if for every prime ideal P of R, and every element $c \in C(P)$, there is a neighborhood U of P (with respect to the patch topology) such that if $Q \in X \cap U$ then $c \in C(Q)$.

This definition was suggested by K.R. Goodearl, and helps to clarify the discussion of various continuity results on Spec (R). An example where generic regularity fails is in [22].

Lemma 1. If R is a Noetherian ring and X a subset of Spec(R), then X has the generic regularity property if either (i) R is FBN, (ii) for every $Q \in X$,

R/Q is a domain, or (iii) there is a uniform bound on the Goldie ranks of the rings R/Q, $Q \in X$.

Proof. If R is FBN, and $c \in C(P)$, then there is an ideal I properly containing P such that $c(R/P) \supseteq I + P/P$. This ideal is easily shown to satisfy our condition (as in the proof of [6, Lemma 10]. For (ii), we use the fact that if $c \in R$ and $Q \in X$, then either $c \in Q$ or $c \in C(Q)$, so that we can let $I = \cap \{Q \in X: c \in Q\}$. Of course, (iii) is a generalization of (ii), but we have singled out (ii) because it arises in practice and is much easier to prove. Part (iii) is an immediate consequence of Stafford's continuity theorem, proved in [21]. (Essentially this connection is made in [23].

If A is a finitely generated right R-module and P is a prime ideal of R such that R/P is right Goldie, then A/AP is an R/P-module. If $Q(R/P)$ is the right (Goldie) quotient ring of R/P, then $(A/AP) \otimes Q(R/P)$ is a semisimple module over the Artinian simple ring $Q(R/P)$. We let

$u_P(A)$ = the number of summands in a decomposition of $(A/AP) \otimes Q(R/P)$ as a direct sum of simple modules.

This is sometimes called the ***reduced rank*** of A/AP. When considering families of prime ideals, it becomes essential to normalize, so we let

$$r_P(A) = u_P(A)/u_P(R_R) .$$

(Note that $r_P(A)$ is closely related to the number of generators of the module $(A/AP) \otimes Q(R/P)$, usually written $g(A,P)$. One easily checks that $g(A,P) = -[-r_P(A)]$.)

Before proceeding, we need a lemma on the behavior of modules over a prime ring, which is essentially a version of [12, Lemma 4.6].

Lemma 2. Let R be a prime right Goldie ring, A and B right R-modules of the same finite torsion-free rank, and assume that A is torsionless. Then A

is isomorphic to a submodule A' of B and B/A' is singular. (In particular, if B is torsion-free then A' is an essential submodule of B.)

Remark: In applications one frequently makes use of the fact that all finitely generated torsion-free right R-modules are torsionless if and only if R is left Goldie.

Proof. Let U be a uniform right ideal of R. If V is any nonzero right ideal of R, then $UV \neq 0$, and $UV \subseteq U$, and it follows immediately from this that for any torsionless module A, $Hom(A,U)$ separates points in A. Let k be the torsion-free rank of A. We choose inductively elements $f_i \in Hom(A,U)$ such that $f_i \neq 0$ and $ker(f_1) \cap \cdots \cap ker(f_{i-1}) \nsubseteq ker(f_i)$. Since the submodules $ker(f_i)$ are closed submodules of A, this process stops when $i = k$. We thus get an embedding $f : A \to \oplus_{i=1}^{k} U$.

Now since B has torsion-free rank k, B contains a submodule isomorphic to $\oplus_{i=1}^{k} V_i$ where each V_i is a uniform right ideal of R. Since each V_i has a submodule isomorphic to U (because $V_i U \neq 0$), we finally get an embedding $A \to B$, which clearly has the indicated properties.

Lemma 3. Let R be a Noetherian ring and X a subset of Spec(R) satisfying the generic regularity condition. Let P be a prime ideal of R and A a finitely generated R-module. Then there is a neighborhood U of P with respect to the patch topology such that if $Q \in U \cap X$ then $r_Q(A) = r_P(A)$.

Proof. First assume that $r_P(A) = 0$, so that A/AP is a torsion R/P-module. Choose a finite set Y of generators for A. We can find an element $c \in C(P)$ such that $yc \in AP$ for all $y \in Y$. According to the generic regularity condition, there is a neighborhood U of P such that if $Q \in X \cap U$ then $c \in C(Q)$. It follows that if $Q \in X \cap U$, then for all $y \in Y$, $yc \in AP \subseteq AQ$, and

$c \in C(Q)$, so A/AQ is a torsion R/Q-module. Hence $r_Q(A) = 0$.

We now assume that $r_P(A) > 0$. Replacing A by A^n for some positive integer n if necessary, we may assume that there is a free R/P-module F such that $r_P(A) = r_P(F)$. Lemma 2 implies that there are exact sequences

$$0 \to F \to A/AP \to C \to 0$$

and

$$A/AP \to F \to D \to 0$$

in which both C and D are torsion as R/P-modules. (Here we use the fact that finitely generated torsion-free R/P-modules are torsionless.) From what we have proved above, there is a neighborhood U of P with respect to the patch topology such that if $Q \in U \cap X$ then $r_Q(C) = r_Q(D) = 0$. It follows that for $Q \in U \cap X$, $r_Q(A) = r_Q(F)$. Since $r_Q(F) = r_P(F) = r_P(A)$, the lemma is proved.

The next two lemmas allow us to choose elements regular at a family of prime ideals (or homomorphisms with good properties at these primes) where we use cardinality considerations rather than the topology of Spec(R). The lemmas and their proofs were suggested by [16, 8.42].

Lemma 4. Let R be a semi-simple Artinian ring, A and B R-modules with B finitely generated, and K a submodule of A. Let f and g be elements of $Hom(A,B)$ such that either $f(K) = B$ or $g(K) = B$. If a and b are elements of the center of R, then $(af + bg)(K) = B$ for all but a finite number of choices of $a^{-1}b$.

Proof. Suppose $g(K) = B$. Let $\varphi: B \to A$ be a splitting for g with $\varphi(B) \supseteq K$. Set $h = f\varphi$ and $d = a^{-1}b$. Let

$$C_d = \{x \in B: (h+d)x = 0\} .$$

As in [16, 8.42], one easily sees that the C_d are independent submodules, so

$C_d = 0$ for all but finitely many choices of d. If $C_d = 0$, then we show that $(af + bg)(K) = B$ by showing that $(af + bg)$ is one-to-one when restricted to $\varphi(B)$. If $x \in B$, and $(af + bg)(\varphi(x)) = 0$, then

$$0 = a^{-1}(af + bg)(\varphi(x)) = f\varphi(x) + a^{-1}bx = (h+d)x ,$$

so $x \in C_d$. Hence, if $C_d = 0$, then $(af + bg)(K) = B$.

If R is a Noetherian ring and P is a prime ideal of R, then for every homomorphism $f: A \to B$ between R-modules, there is an induced homomorphism which we denote f_P:

$$f_P: (A/AP) \otimes Q(R/P) \to (B/BP) \otimes Q(R/P) .$$

We say that f is an *epimorphism at* P if f_P is an epimorphism. We may, of course, identify $(B/BP) \otimes Q(R/P)$ with $B \otimes Q(R/P)$, and we note that if P is a localizable prime ideal, and B finitely generated, then f is an epimorphism at P if and only if the induced map of localizations $A_P \to B_P$ is an epimorphism.

Lemma 5. Let R be a Noetherian ring and X an infinite set of prime ideals of R. Let A and B be R-modules, with B finitely generated, and for each $P \in X$, let K_P be a $Q(R/P)$-submodule of $A \otimes Q(R/P)$. Let f and g be elements of $Hom(A,B)$ such that for $P \in X$, either $f_P K_P = B \otimes Q(R/P)$ or $g_P K_P = B \otimes Q(R/P)$. Suppose that R contains a set F of central units such that if α and β are distinct elements of F, then $\alpha - \beta$ is a unit, and such that $|X| < |F|$. Then there are central units α and β with $\alpha - \beta = 1$ and such that $(\alpha f_P - \beta g_P)(K_P) = B \otimes Q(R/P)$ for all $P \in X$.

Proof. For every prime P, the natural map of F into the center of $Q(R/P)$ is injective, since the difference of two distinct elements of F is a unit. We conclude from Lemma 4 that there are elements α and β of F such that $\alpha \neq \beta$ and $(\alpha f_P - \beta g_P)(K_P) = B \otimes Q(R/P)$ for all $P \in X$. Since $\alpha - \beta$ is also a unit, we may multiply by $(\alpha-\beta)^{-1}$, thus obtaining central units α and β (not necessarily in F) such that $\alpha-\beta=1$ and $(\alpha f_P-\beta g_P)(K_P)=B\otimes Q(R/P)$ for all $P\in X$.

If X is a subset of Spec(R), we let $C(X)=\cap\{C(P): P\in X\}$. Thus if X has the generic regularity property, and $c\in C(Q)$, then there is a neighborhood U of Q (with respect to the patch topology) such that $c\in C(X\cap U)$. We say that a nonempty set X of prime ideals in R satisfies the *intersection condition* if every one-sided ideal of R which has nonempty intersection with $C(P)$ for all $P\in X$ also has nonempty intersection with $C(X)$.

Lemma 6. Let R be a Noetherian ring and let X be a subset of Spec(R) with the generic regularity property. Assume that R contains a set F of central units such that the difference of any two distinct elements of F is a unit, and such that $|X|<|F|$. Then X satisfies the intersection condition.

Proof. Let X-Spec(R) be the set of primes which are intersections of primes in the set X. Clearly, X-Spec(R) is a closed subset of Spec(R) with respect to the patch topology. Let I be a right ideal of R such that $I\cap C(P)\neq\varphi$ for all $P\in X$. If $Q\in X$-Spec(R), and $I'=I+Q/Q$, then the left annihilator of I' is contained in P for every $P\in X\cap V(Q)$, since I contains an element of $C(P)$. Hence, since the intersection of these primes is Q, we see that I' has zero left annihilator and thus is essential in R/Q. Thus $I\cap C(Q)\neq\phi$. If $c\in I\cap C(Q)$, then the generic regularity condition implies that there is a neighborhood U of Q with respect to the patch topology such that if $P\in X\cap U$ then $c\in C(P)$. We have thus a covering of X-Spec(R) by open sets. We select a finite subcover, say $\{U_1,\cdots,U_m\}$ and note that there are elements $c_1,\cdots,c_m$ of I such that $c_i\in(U_i\cap X)$. We now apply Lemma 5 in the situation in which $A=B=R$, and for each $P\in X$, $K_P=R/P$, and in which the homomorphisms are left multiplication by the elements c_i, $i = 1, \cdots, m$. Applying Lemma 5 several times, we find that there are central units $\alpha_1, \cdots, \alpha_m$ such that if $c = \alpha_1 c_1 + \cdots + \alpha_m c_m$, then $c \in I \cap C(X)$.

2. Localization.

If R is a ring and C a multiplicatively closed subset, then we can ask for a ring T and a ring homomorphism $u: R \to T$ such that the elements in $u(C)$ are units in T and the pair (u, T) is universal with respect to this property. Such a construction can clearly always be made. We say that C is a *right denominator set* if (i) every element of T is of the form $u(r)u(c)^{-1}$ for some $r \in R$, $c \in C$, and (ii) $u(s) = 0$ if and only if for some $c \in C$, $sc = 0$. It is well known (cf. [8, 12.1.2]) that C is a right denominator set if and only if (i) for every $r \in R$ and $c \in C$, there are elements $s \in R$ and $d \in C$ with $rd = cs$ (the *right Ore condition)* and (ii) for every $r \in R$ and $c \in C$, if $cr = 0$ then for some $d \in C$, $rd = 0$ (the *right reversibility condition).* It is also well known that if R is Noetherian (more generally, if R satisfies the ascending chain condition on right annihilators of elements) that the right Ore condition implies the right reversibility condition ([8, 12.4.4]). We say a prime ideal P in a Noetherian ring R is *localizable* if $C(P)$ is a left and right Ore set. In general, however, $C(P)$ is not an Ore set, and we attempt instead to find an Ore set contained in $C(P)$ and as large as possible.

Definition. If P and Q are prime ideals of the Noetherian ring R, there is a *second layer link* $Q \quad P$ if there is an ideal J, with $Q \cap P \supset J \supseteq QP$, such that $J \neq Q \cap P$ and $Q \cap P / J$ is torsion-free as a left R/Q-module and as a right R/P-module. (This notion originates in [14].) The *graph of links* of R is the directed graph whose vertices are the elements of Spec(R) and whose arrows are given by the second layer links. A *clique* in Spec(R) is a connected component of the graph of links, [16]. If $P \in \operatorname{Spec}(R)$, we let $Cl(P)$ be the unique clique containing P.

Theorem 7. If R is a Noetherian ring, then every clique in $\operatorname{Spec}(R)$ is finite or countable.

A proof is outlined in [23].

It is an easy consequence of the reversibility condition that if S is an Ore set in a Noetherian ring R and $S \subseteq C(P)$, then $S \subseteq C(Q)$ for all $Q \in Cl(P)$. This leads us to a candidate for the largest Ore set in $C(P)$. If X is a clique in $\operatorname{Spec}(R)$, then we let $C(X) = \cap\{C(P): P \in X\}$. The clique X is *classical* if (i) $C(X)$ is a right and left Ore set, so that a localization $R_X = R\,C(X)^{-1}$ exists, (ii) for every prime Q, $Q \in X$, R_X/QR_X is naturally isomorphic to the Goldie quotient ring of R/Q. (iii) the primes QR_X $Q \in X$, are precisely the primitive ideals of R_X, and (iv) the R_X-injective hull of every simple R_X-module is the union of its socle sequence.

For such a localization to exist, there is a necessary ideal theoretic condition, first noted by Jategaonkar in [15].

Definition. A prime P in a Noetherian ring R satisfies the *second layer condition* if the injective hull of R/P contains no finitely generated submodule whose annihilator is a prime ideal other than P.

It is easy to see that every prime ideal in a fully bounded Noetherian ring satisfies this condition. K. Brown and A. Jategaonkar have shown independently that if R is the group ring of a polycyclic-by-finite group or the enveloping algebra of a solvable Lie algebra over an algebraically closed field of characteristic zero, then all primes in R satisfy this condition, [4][15][16]. These results have been extended to certain iterated differential operator rings, skew group rings and crossed products by A. Bell [2]. On the other hand, the co-finite dimensional maximal ideals of the universal enveloping algebras of simple Lie algebras always fail to satisfy the second layer condition.

Theorem 8. Let R be a Noetherian ring whose prime ideal satisfy the second layer condition and which contains an uncountable set F of central units such that the difference of two distinct elements of F is a unit. Let X be a clique in Spec(R) which satisfies the generic regularity condition. Then X is classical.

Proof. Lemma 6 and Theorem 7 together show that the clique X satisfies the intersection condition. The proofs of [16, 8.35] or [5, 3.5] now apply to prove the result.

Remark. This result was obtained independently by J.T. Stafford and the author in June 1983. A slightly improved version is stated in [23].

The object of a localization theory is to be able to prove theorems which do not themselves refer to localization. As an example, we note the result in [6] that if R is a Noetherian ring in which all cliques are classical, then the global dimension of R (if it is finite) is bounded below by the classical Krull dimension of R. The examples mentioned above give a wide variety of rings to which this applies. One can also study the global dimension of a fully bounded ring by refering to its localizations, as the following lemma indicates.

Lemma 9. Let R be a fully bounded Noetherian ring in which all cliques are classical. Then

$$\text{gl. dim. } R = \mathit{sup}\ \{ \text{gl. dim. } R_X \}$$

where the supremum is taken over all cliques X consisting of maximal ideals.

This is essentially [18, Theorem 7, and Corollary 8].

3. Local Algebra.

The purpose of this section is to show that the rings arising from localizations at infinite cliques do satisfy many of the properties one would like in localized rings, even though they are not local or semilocal. We show, in effect, that essentially all of the good properties of localizations of rings which are module-finite over their centers extend to this situation. (For example, when we localize a commutative ring R at a prime P, the resulting local ring R_P has a stronger property than the one in Theorem 10 (vi) below, since we have a theory of projective covers. However, if R is a commutative ring and A an R-algebra, finitely generated as an R-module, and again we localize at a prime P of R, obtaining a localized algebra A_P, then A_P is semilocal, but we do not in general have projective covers, and one can generally not do better than the result of Theorem 10, (vi) below.) The natural way to think about the results in this section is that they show that the localized rings have a zero-dimensional K-theory. That is, the results are suggestively similar to those in [11] and to the (trivial) zero-dimensional special cases of K-theoretic results for fully bounded rings, as in [25] and elsewhere.

Theorem 10. Let R be a Noetherian ring and X the set of primitive ideals of R, and assume that (i) for $P \in X$, R/P is Artinian, (ii) either X is finite or R contains a set F of central units such that the difference of two distinct elements of F is a unit, and such that $|F| > |X|$, and (iii) X satisfies the generic regularity condition. Then

(i) If A and B are R-modules, with B finitely generated, and if for every J-prime P there is a homomorphism $f: A \to B$ such that the induced map $f': A \otimes Q(R/P) \to B \otimes Q(R/P)$ is an epimorphism (equivalently, $r_P(B/f(A)) = 0$), then there is an epimorphism from A to B.

(ii) If A and B are R-modules with B finitely generated and A projective, then there is an epimorphism from A to B if and only if for each $P \in X$, $r_P(A) \geqq r_P(B)$.

(iii) If A and B are finitely generated projective modules, then $A \cong B$ if and only if for all $P \in X$, $r_P(A) = r_P(B)$.

(iv) The Bass stable rank of R is one.

(v) If A is a finitely generated projective R-module and B and C are arbitrary R-modules such that $A \oplus B \cong A \oplus C$, then $B \cong C$.

(vi) If A is a finitely generated projective module and B is a module and f and g are epimorphisms from A to B, then there is an automorphism φ of A such that $f = g\varphi$.

Remarks: (i) This theorem seems to be unknown even in the commutative case, presumably because the condition that there are more units than maximal ideals is not natural there.

(ii) When X is finite, all of these results are well known. There are references in the various parts of the proof.

(iii) Result (i) can be thought of as an analogue of Serre's theorem on free summands. In the commutative case the zero-dimensional version of Serre's theorem would say that if a module locally has R as a summand, then globally it has R as a summand. The local condition that A_P has R_P as a summand just means that there is a homomorphism $f: A \to R$ such that the induced map $A_P/PA_P \to R_P/RP_P$ is an epimorphism, and this is precisely our condition.

Proof. To prove (i), let $P \in J$-Spec(R) and choose a homomorphism $f: A \to B$ such that f is an epimorphism at P -- that is, the induced map $A \otimes Q(R/P) \to B \otimes Q(R/P)$ is an epimorphism, or, equivalently, $r_P(B/f(A)) = 0$. According to Lemma 3 there is a neighborhood U of P, with respect to the patch topology, such that for $Q \in X \cap U$, $r_Q(B/f(A)) = 0$. Using compactness, we obtain a covering of J-Spec(R) by open sets $U_1, \cdots, U_m$ and homomorphisms $f_1, \cdots, f_m$ from A to B such that if $Q \in X \cap U_i$ then $r_Q(B/f_i(A)) = 0$. Applying Lemma 5, we find central units $\alpha_1, \cdots, \alpha_m$ such that if $f = \alpha_1 f_1 + \cdots + \alpha_m f_m$ then for all $Q \in X$, $r_Q(B/f(A)) = 0$. This implies that f is an epimorphism.

We prove (ii) as an application of (i), and we thus must show that for each J-prime T, there is a homomorphism $f: A \to B$ such that the induced map $f': A \otimes Q(R/T) \to B \otimes Q(R/T)$ is an epimorphism. Since A is projective, it suffices to find a homomorphism from A/AT to B/BT whose cokernel is singular as an R/T-module. To do this, it suffices to know that $r_T(A) \geq r_T(B)$, (using Lemma 2). According to Lemma 3, there is a neighborhood U of T in the patch topology such that if $P \in X \cap U$ then $r_P(A) = r_T(A)$ and $r_P(B) = r_T(B)$. Since $r_P(A) \geq r_P(B)$ by hypothesis, the result follows.

Clearly, (iii) is an immediate consequence of (ii).

To say that the stable rank of R is one means that if $a_1 x_1 + \cdots + a_n x_n = 1$ then for suitable elements $y_2, \cdots, y_n$ in R, $a_1 + a_2 y_2 + \cdots + a_n y_n$ is a unit. (For the general setting, left-right symmetry, and some consequences we refer to [26].) This is well known for semi-local rings ([1], Lemma 6.4]) so we may assume that the set X of maximal ideals is infinite. If P is any prime ideal of R then by [20, Lemma 1.1] or [25, Lemma 2], there are elements $s_2, \cdots, s_n$ of R such that if

$$w' = a_1 + a_2 s_2 + \cdots + a_n s_n ,$$

then $w' \in C(P)$. The generic regularity condition implies that there is a neighborhood U of P in the patch topology such that if $Q \in X \cap U$, then $w' \in C(Q)$. Doing this for every prime P, we obtain a covering of Spec(R) by open sets with this property. Since Spec(R) is compact with respect to the patch topology, we may choose a finite subcover, and thus we obtain subsets X_i, $(i = 1, \cdots, m)$ of X such that $\cup_{i=1}^m X_i = X$ and for each index i, there is an element w_i and elements s_{ji} $(j = 1, \cdots, n)$ such that

$$w_i = a_1 + \Sigma_{j=1}^n a_j s_{ji}$$

and $w_i \in C(Q)$ for all $Q \in X_i$. We now apply Lemma 5 several times, in the special case in which $R = A = B$ and the homomorphisms considered are left multiplications by the elements w_i, and we obtain an element

$$w = \alpha_1 w_1 + \cdots + \alpha_m w_m$$

where the α_i are all units, $\Sigma_{i=1}^m \alpha_i = 1$, and $w \in C(Q)$ for all $Q \in X$. This implies that w is a unit, and completes the proof of (v).

That (vi) is a consequence of (v) was proved in [9]. Statement (vii) is a generalization of (vi) and follows from [24, Theorem 7].

It is clear that part (iii) of Theorem 10 should have consequences for the structure of $K_0(R)$, but we are only able to obtain a nice result with an additional countability hypothesis (satisfied in our main examples).

Theorem 11. Let R be a Noetherian ring satisfying the hypothesis of Theorem 10 and assume in addition that the set X of primitive ideals is countable. Then $K_0(R)$ is free Abelian of at most countable rank.

Proof. If J is the Jacobson radical of R, then the natural map $K_0(R) \to K_0(R/J)$ is injective, so we may assume that $J = 0$. Also, if $0 = P_1 \cap \cdots \cap P_n$, then the natural map $K_0(R) \to \Pi_{i=1}^n K_0(R/P_i)$ is injective (using Theorem 10, part (iii)), so we may assume that R is a prime ring whose Jacobson radical is zero. By Noetherian induction, we may assume that the result holds for any proper homomorphic image of R.

For every prime P, there is a homomorphism

$$s_P : K_0(R) \to \mathbb{Q}$$

defined by $s_P([A]-[B]) = r_P(A) - r_P(B)$. It is clear that $s_P(K_0(R))$ is a cyclic subgroup of $\mathbb{Q}$, since $r_P(A)$ can always be written as a fraction whose denominator is the Goldie rank of R/P. (These homomorphisms are exploited in [12].) Part (iii) of Theorem 10 implies that if x and y are elements of $K_0(R)$ and $s_P(x) = s_P(y)$ for all $P \in X$, then $x = y$.

Since (0) is a prime ideal, we have a homomorphism $s_{(0)}: K_0(R) \to \mathbb{Q}$, whose image is cyclic, and it will suffice to show that the kernel of this map is a free Abelian group. We call this kernel K'. For every nonzero ideal I of R, we define a subgroup K_I of $K_0(R)$ by $K_I = \{ x \in K' : s_P(x) = 0 \text{ if } P \in W(I) \}$. Clearly, the natural map $K_0(R) \to K_0(R/I)$ carries K_I injectively into $K_0(R/I)$, and so, by Noetherian induction, K_I is free and countable. Now if x is any element of K', the generic regularity condition and Lemma 3 imply that there is a neighborhood U of (0) in the patch topology such that $s_P(x) = 0$ for $P \in X \cap U$. Since the neighborhoods of (0) are all of the form $W(I)$, for nonzero ideals I, this means that K' is the union of the free subgroups K_I. We further note that these are pure subgroups -- that is, if for some nonzero integer n, $nx \in K_I$, then $x \in K_I$.

If $\overline{I} = \cap\{P : P \in X, P \geq I\}$, then $K_I = K_{\overline{I}}$, so we need only consider radical ideals I, i.e. ideals I with $I = \overline{I}$. We choose a descending cofinal system of radical ideals I indexed by ordinals. Since X is countable, this is a countable cofinal system, so there is actually a sequence $\{I_i : 1 \leq i \leq \infty\}$ of radical ideals, such that $K' = \cup_{i=1}^{\infty} K_{I_i}$. An old result of Pontriagin's [19] shows that a group which is an ascending union of a sequence of countable, pure, free subgroups is itself free. This proves Theorem 11.

We now turn to our results on global dimension.

Theorem 12. Let R be a Noetherian ring satisfying the hypotheses of Theorem 10. Then for every finitely generated R-module A, the following three conditions are equivalent:

(i) pr. dim $A \leq n$,

(ii) $Ext_R^{n+1}(A,S) = 0$ for all simple right modules S, and

(iii) $Tor_R^{n+1}(A,T) = 0$ for all simple left modules T.

In particular, gl. dim. $R = sup$ { pr. dim. S: S a simple right module } = sup { inj. dim. S: S a simple right module }.

Remark. Note that there is no assumption of finite global dimension to begin with. It is well known ([3], [10, Theorem [6]) that if the global dimension of a Noetherian ring R is finite, then gl. dim. $R = sup$ { pr. dim. S: S a simple right module }. In general, however, it is not known whether the simple modules can give a criterion for the global dimension to be finite. The formula involving injective dimension is, of course, false for many Noetherian rings, including hereditary ones (in which all simples may be injective.)

Proof. We may replace S and T by modules of the form R/M where M is a maximal, co-Artinian ideal. Since R/M is simple Artinian, it is an injective cogenerator in the category of R/M-modules. Thus [7, Prop. 5.3, p. 120] implies that

$$Ext_R^n(A,R/M) = Hom(Tor_R^n(A,R/M), R/M) .$$

(An easy direct proof of this can be given using the fact that for every right R-module B, $B \otimes (R/M) \cong B/BM$.) In particular, $Ext_R^{n+1}(A,R/M) = 0$ if and only if $Tor_R^{n+1}(A,R/M) = 0$. This shows the equivalence of (ii) and (iii).

It is also clear that (i) implies (ii) and (iii). Assuming that (iii) holds, choose a projective resolution of A:

$$0 \to K_n \to P_{n-1} \to \cdots \to P_0 \to A \to 0$$

where the P_i are all projective, but K_n may not be, and a resolution for K_n:

$$0 \to K_{n+1} \to P_n \to K_n \to 0.$$

Now $Tor_R^{n+1}(A,R/M) = Tor_R^1(K_n,R/M) = 0$ so the induced map $K_{n+1} \otimes R/M \to P_n \otimes R/M$ is injective. Since for any module B, $B \otimes (R/M) = B/BM$, we conclude that for every maximal ideal M of R, $K_{n+1}M = P_nM \cap K_{n+1}$. The proof will be complete if we can conclude from this that K_{n+1} is a summand of P_n, which is precisely what the following lemma will enable us to do.

Lemma 13. Let R be a Noetherian ring satisfying the hypotheses of Theorem 10. Let A be a finitely generated projective module and K a submodule such that for all $P \in X$, $KP = K \cap AP$. Then K is a summand of A.

Proof. We consider a projective module A and a submodule K with natural inclusion map $\varphi: K \to A$, and we want to construct a homomorphism $f: A \to K$ such that $f\varphi = 1_K$. Since K is Noetherian, it is sufficient to construct a homorphism $f: A \to K$ such that $f\varphi$ is surjective. To show $f\varphi$ surjective, it suffices to check that for each $P \in X$, $f\varphi(K) + KP = K$. In effect, our hypothesis says that for each $P \in X$ we can choose an f which works for all maximal ideals simultaneously. As in the proof of (ii), we first find such an f for each J-prime ideal. If $T \in J$-Spec(R), then $T = \cap\{P \in X: P \geq T\}$, so if B is a torsion-free right R/T-module, then $\cap\{BP: P \in X, P \geq T\} = 0$. It follows that the induced map $\varphi': K \otimes Q(R/T) \to A \otimes Q(R/T)$ is injective, and hence splits. We let $g: A \otimes Q(R/T) \to K \otimes Q(R/T)$ be the splitting. Thus $g(A/AT)$ and

$K/(K\cap AT)$ are both finitely generated torsionless, essential R/T-submodules of $K\otimes Q(R/T)$. Lemma 2 implies that $g(A/AT)$ is isomorphic to an essential submodule of $K/(K\cap AT)$. Using the projectivity of A, we conclude that there is a homomorphism $g_1: A\to K$ such that $g_1\varphi$ is a epimorphism at T (i.e. $r_T(K/g_1\varphi(K))=0$.) It follows from Lemma 3 that there is a neighborhood U of T in the patch topology such that if $P\in X\cap U$, then $r_P(K/g_1\varphi(K))=0$. Proceeding as in the proof of Theorem 10, (i), we find a finite set $\{U_1,\cdots,U_n\}$ of open sets covering J-Spec(R), and homomorphisms $g_i: A\to K$ such that for $P\in X\cap U_i$, $r_P(K/g_i\varphi(K))=0$. Patching these together, as in the proof of Theorem 10, we find a homomorphism $f=\alpha_1 g_1+\cdots+\alpha_n g_n$ such that $f\varphi$ is an epimorphism at every maximal ideal, and hence is an actual epimorphism.

Corollary 14. Let R be a fully bounded Noetherian ring with an uncountable set F of central units such that the difference of two elements of F is a unit. Then

$$\text{gl. dim. } R = sup\,\{\text{ pr. dim. } S: S \text{ a simple } R\text{-module }\}.$$

Proof. We can apply Theorem 8 to localize R at cliques of maximal ideals, since Lemma 1 shows that cliques in Spec(R) satisfy the generic regularity condition, and the second layer condition for FBN rings is routine. Lemma 9 shows that gl. dim. $R = sup\,\{$ gl. dim. $R_X\}$, where the supremum if taken over cliques of maximal ideals, and Theorem 12 shows that gl. dim. R_X is the supremum of the projective dimensions of simple R_X-modules. Since all simple R_X-modules are induced from simple R-modules, this proves Corollary 14.

Note added August 1984: Since this paper was written, L. Small and J.T. Stafford and (independently) K.R. Goodearl have given alternative proofs of

Corollary 14 which show that the result holds for all fully bounded Noetherian rings.

REFERENCES

1. H. Bass, K-theory and stable algebra, *Pub. Math. I.H.E.S.* **22** (1964), 5-60.

2. A. Bell, Localization and ideal theory in Noetherian crossed products and differential operator rings, dissertation, University of Washington, 1984.

3. S.M. Bhatwadekar, On the global dimension of some filtered algebras, *J. London Math. Soc. (2)* **13** (1976), 239-248.

4. K.A. Brown, Module extensions over Noetherian rings. *J. Algebra* **69** (1981), 247-260.

5. K.A. Brown, Ore sets in Noetherian rings, to appear in *Séminaire Malliavin.*

6. K.A. Brown and R.B. Warfield, Jr., Krull and global dimensions of fully bounded Noetherian rings, *Proc. Amer. Math. Soc.*, (to appear).

7. H. Cartan and S. Eilenberg, "Homological Algebra", Princeton, 1956.

8. P.M. Cohn, "Algebra", vol. 2, London, Wiley, 1977.

9. L. Fuchs, On a substitution property of modules, *Monatshefte f. Math.* **75**(1971), 198-204.

10. K.R. Goodearl, Global dimension of differential operator rings, II., *Trans. Amer. Math. Soc.* **209**(1975), 65-85.

11. K.R. Goodearl and R.B. Warfield, Jr., Algebras over zero-dimensional rings, *Math. Ann.* **223**(1976), 157-168.

12. K.R. Goodearl and R.B. Warfield, Jr., State spaces of K_0 of Noetherian rings, *J. Alg.* **71**(1981), 322-378.

13. M. Hochster, Prime ideal structure in commutative rings, *Trans. Amer. Math. Soc.* **142**(1969), 43-60.

14. A.V. Jategaonkar, Injective modules and classical localization in Noetherian rings, *Bull. Amer. Math. Soc.* **79**(1973), 152-157.

15. A.V. Jategaonkar, Solvable Lie algebra, polycyclic-by-finite groups and bimodule Krull dimension, *Comm. Alg.* **10**(1982), 19-69.

16. A.V. Jategaonkar, Localization in Noetherian rings, (1983 preliminary version).

17. T.H. Lenagan, Artinian ideals in Noetherian rings, *Proc. Amer. Math. Soc.* **51**(1975), 499-500.

8. B.J. Mueller, Links between maximal ideals in bounded Noetherian prime rings of Krull dimension one, to appear in proceedings of 1983 Antwerp conference.

9. L. Pontriagin, The theory of topological commutative groups, *Ann. Math.* **35**(1934), 361-388.

0. J.T. Stafford, Stable structure of noncommutative Noetherian rings, *J. Algebra* **47**(1977), 244-267.

21. J.T. Stafford, Generating modules efficiently: algebriac K-theory for non-commutative Noetherian rings, *J. Algebra* **69**(1981), 312-346; (orig. ibid. 82 (1983), 294-296).

22. J.T. Stafford, On the ideals of a Noetherian ring, (preprint).

23. J.T. Stafford, The Goldie rank of a module, to appear in the proceedings of the 1983 Oberwolfach conference on Noetherian rings.

24. R.B. Warfield, Jr., Modules over fully bounded rings, in "Ring Theory, Waterloo, 1978" (proceedings), ed. D. Handelman and J. Lawrence, *Lecture Notes in Math.* **734**, Springer-Verlag, Berlin, 1979, pp. 339-352.

25. R.B. Warfield, Jr., Stable generation of modules, in "Module Theory", ed. C. Faith and S. Wiegand, *Lecture Notes in Mathematics* **700**, Springer-Verlag, Berlin, 1979, pp. 16-33.

26. R.B. Warfield, Jr., Cancellation of modules and groups and stable range of endomorphism rings, *Pacific J. Math.* **91**(1980), 457-485.

Robert B. Warfield, Jr.
Department of Mathematics
University of Washington
Seattle, Washington 98195
U.S.A.

Vol. 1062: J. Jost, Harmonic Maps Between Surfaces. X, 133 pages. 1984.

Vol. 1063: Orienting Polymers. Proceedings, 1983. Edited by J.L. Ericksen. VII, 166 pages. 1984.

Vol. 1064: Probability Measures on Groups VII. Proceedings, 1983. Edited by H. Heyer. X, 588 pages. 1984.

Vol. 1065: A. Cuyt, Padé Approximants for Operators: Theory and Applications. IX, 138 pages. 1984.

Vol. 1066: Numerical Analysis. Proceedings, 1983. Edited by D.F. Griffiths. XI, 275 pages. 1984.

Vol. 1067: Yasuo Okuyama, Absolute Summability of Fourier Series and Orthogonal Series. VI, 118 pages. 1984.

Vol. 1068: Number Theory, Noordwijkerhout 1983. Proceedings. Edited by H. Jager. V, 296 pages. 1984.

Vol. 1069: M. Kreck, Bordism of Diffeomorphisms and Related Topics. III, 144 pages. 1984.

Vol. 1070: Interpolation Spaces and Allied Topics in Analysis. Proceedings, 1983. Edited by M. Cwikel and J. Peetre. III, 239 pages. 1984.

Vol. 1071: Padé Approximation and its Applications, Bad Honnef 1983. Prodeedings. Edited by H. Werner and H.J. Bünger. VI, 264 pages. 1984.

Vol. 1072: F. Rothe, Global Solutions of Reaction-Diffusion Systems. V, 216 pages. 1984.

Vol. 1073: Graph Theory, Singapore 1983. Proceedings. Edited by K.M. Koh and H.P. Yap. XIII, 335 pages. 1984.

Vol. 1074: E.W. Stredulinsky, Weighted Inequalities and Degenerate Elliptic Partial Differential Equations. III, 143 pages. 1984.

Vol. 1075: H. Majima, Asymptotic Analysis for Integrable Connections with Irregular Singular Points. IX, 159 pages. 1984.

Vol. 1076: Infinite-Dimensional Systems. Proceedings, 1983. Edited by F. Kappel and W. Schappacher. VII, 278 pages. 1984.

Vol. 1077: Lie Group Representations III. Proceedings, 1982–1983. Edited by R. Herb, R. Johnson, R. Lipsman, J. Rosenberg. XI, 454 pages. 1984.

Vol. 1078: A.J.E.M. Janssen, P. van der Steen, Integration Theory. V, 224 pages. 1984.

Vol. 1079: W. Ruppert. Compact Semitopological Semigroups: An Intrinsic Theory. V, 260 pages. 1984

Vol. 1080: Probability Theory on Vector Spaces III. Proceedings, 1983. Edited by D. Szynal and A. Weron. V, 373 pages. 1984.

Vol. 1081: D. Benson, Modular Representation Theory: New Trends and Methods. XI, 231 pages. 1984.

Vol. 1082: C.-G. Schmidt, Arithmetik Abelscher Varietäten mit komplexer Multiplikation. X, 96 Seiten. 1984.

Vol. 1083: D. Bump, Automorphic Forms on GL (3,IR). XI, 184 pages. 1984.

Vol. 1084: D. Kletzing, Structure and Representations of Q-Groups. VI, 290 pages. 1984.

Vol. 1085: G.K. Immink, Asymptotics of Analytic Difference Equations. V, 134 pages. 1984.

Vol. 1086: Sensitivity of Functionals with Applications to Engineering Sciences. Proceedings, 1983. Edited by V. Komkov. V, 130 pages. 1984

Vol. 1087: W. Narkiewicz, Uniform Distribution of Sequences of Integers in Residue Classes. VIII, 125 pages. 1984.

Vol. 1088: A.V. Kakosyan, L.B. Klebanov, J.A. Melamed, Characterization of Distributions by the Method of Intensively Monotone Operators. X, 175 pages. 1984.

Vol. 1089: Measure Theory, Oberwolfach 1983. Proceedings. Edited by D. Kölzow and D. Maharam-Stone. XIII, 327 pages. 1984.

Vol. 1090: Differential Geometry of Submanifolds. Proceedings, 1984. Edited by K. Kenmotsu. VI, 132 pages. 1984.

Vol. 1091: Multifunctions and Integrands. Proceedings, 1983. Edited by G. Salinetti. V, 234 pages. 1984.

Vol. 1092: Complete Intersections. Seminar, 1983. Edited by S. Greco and R. Strano. VII, 299 pages. 1984.

Vol. 1093: A. Prestel, Lectures on Formally Real Fields. XI, 125 pages. 1984.

Vol. 1094: Analyse Complexe. Proceedings, 1983. Edité par E. Amar, R. Gay et Nguyen Thanh Van. IX, 184 pages. 1984.

Vol. 1095: Stochastic Analysis and Applications. Proceedings, 1983. Edited by A. Truman and D. Williams. V, 199 pages. 1984.

Vol. 1096: Théorie du Potentiel. Proceedings, 1983. Edité par G. Mokobodzki et D. Pinchon. IX, 601 pages. 1984.

Vol. 1097: R.M. Dudley, H. Kunita, F. Ledrappier, École d'Éte de Probabilités de Saint-Flour XII – 1982. Edité par P.L. Hennequin. X, 396 pages. 1984.

Vol. 1098: Groups – Korea 1983. Proceedings. Edited by A.C. Kim and B.H. Neumann. VII, 183 pages. 1984.

Vol. 1099: C.M. Ringel, Tame Algebras and Integral Quadratic Forms. XIII, 376 pages. 1984.

Vol. 1100: V. Ivrii, Precise Spectral Asymptotics for Elliptic Operators Acting in Fiberings over Manifolds with Boundary. V, 237 pages. 1984.

Vol. 1101: V. Cossart, J. Giraud, U. Orbanz, Resolution of Surface Singularities. Seminar. VII, 132 pages. 1984.

Vol. 1102: A. Verona, Stratified Mappings – Structure and Triangulability. IX, 160 pages. 1984.

Vol. 1103: Models and Sets. Proceedings, Logic Colloquium, 1983, Part I. Edited by G.H. Müller and M.M. Richter. VIII, 484 pages. 1984.

Vol. 1104: Computation and Proof Theory. Proceedings, Logic Colloquium, 1983, Part II. Edited by M.M. Richter, E. Börger, W. Oberschelp, B. Schinzel and W. Thomas. VIII, 475 pages. 1984.

Vol. 1105: Rational Approximation and Interpolation. Proceedings, 1983. Edited by P.R. Graves-Morris, E.B. Saff and R.S. Varga. XII, 528 pages. 1984.

Vol. 1106: C.T. Chong, Techniques of Admissible Recursion Theory. IX, 214 pages. 1984.

Vol. 1107: Nonlinear Analysis and Optimization. Proceedings, 1982. Edited by C. Vinti. V, 224 pages. 1984.

Vol. 1108: Global Analysis – Studies and Applications I. Edited by Yu.G. Borisovich and Yu.E. Gliklikh. V, 301 pages. 1984.

Vol. 1109: Stochastic Aspects of Classical and Quantum Systems. Proceedings, 1983. Edited by S. Albeverio, P. Combe and M. Sirugue-Collin. IX, 227 pages. 1985.

Vol. 1110: R. Jajte, Strong Limit Theorems in Non-Commutative Probability. VI, 152 pages. 1985.

Vol. 1111: Arbeitstagung Bonn 1984. Proceedings. Edited by F. Hirzebruch, J. Schwermer and S. Suter. V, 481 pages. 1985.

Vol. 1112: Products of Conjugacy Classes in Groups. Edited by Z. Arad and M. Herzog. V, 244 pages. 1985.

Vol. 1113: P. Antosik, C. Swartz, Matrix Methods in Analysis. IV, 114 pages. 1985.

Vol. 1114: Zahlentheoretische Analysis. Seminar. Herausgegeben von E. Hlawka. V, 157 Seiten. 1985.

Vol. 1115: J. Moulin Ollagnier, Ergodic Theory and Statistical Mechanics. VI, 147 pages. 1985.

Vol. 1116: S. Stolz, Hochzusammenhängende Mannigfaltigkeiten und ihre Ränder. XXIII, 134 Seiten. 1985.

Vol. 1117: D.J. Aldous, J.A. Ibragimov, J. Jacod, Ecole d'Été de Probabilités de Saint-Flour XIII – 1983. Édité par P.L. Hennequin. IX, 409 pages. 1985.

Vol. 1118: Grossissements de filtrations: exemples et applications. Seminaire, 1982/83. Edité par Th. Jeulin et M. Yor. V, 315 pages. 1985.

Vol. 1119: Recent Mathematical Methods in Dynamic Programming. Proceedings, 1984. Edited by I. Capuzzo Dolcetta, W.H. Fleming and T. Zolezzi. VI, 202 pages. 1985.

Vol. 1120: K. Jarosz, Perturbations of Banach Algebras. V, 118 pages. 1985.

Vol. 1121: Singularities and Constructive Methods for Their Treatment. Proceedings, 1983. Edited by P. Grisvard, W. Wendland and J.R. Whiteman. IX, 346 pages. 1985.

Vol. 1122: Number Theory. Proceedings, 1984. Edited by K. Alladi. VII, 217 pages. 1985.

Vol. 1123: Séminaire de Probabilités XIX 1983/84. Proceedings. Edité par J. Azéma et M. Yor. IV, 504 pages. 1985.

Vol. 1124: Algebraic Geometry, Sitges (Barcelona) 1983. Proceedings. Edited by E. Casas-Alvero, G.E. Welters and S. Xambó-Descamps. XI, 416 pages. 1985.

Vol. 1125: Dynamical Systems and Bifurcations. Proceedings, 1984. Edited by B.L.J. Braaksma, H.W. Broer and F. Takens. V, 129 pages. 1985.

Vol. 1126: Algebraic and Geometric Topology. Proceedings, 1983. Edited by A. Ranicki, N. Levitt and F. Quinn. V, 523 pages. 1985.

Vol. 1127: Numerical Methods in Fluid Dynamics. Seminar. Edited by F. Brezzi, VII, 333 pages. 1985.

Vol. 1128: J. Elschner, Singular Ordinary Differential Operators and Pseudodifferential Equations. 200 pages. 1985.

Vol. 1129: Numerical Analysis, Lancaster 1984. Proceedings. Edited by P.R. Turner. XIV, 179 pages. 1985.

Vol. 1130: Methods in Mathematical Logic. Proceedings, 1983. Edited by C.A. Di Prisco. VII, 407 pages. 1985.

Vol. 1131: K. Sundaresan, S. Swaminathan, Geometry and Nonlinear Analysis in Banach Spaces. III, 116 pages. 1985.

Vol. 1132: Operator Algebras and their Connections with Topology and Ergodic Theory. Proceedings, 1983. Edited by H. Araki, C.C. Moore, Ş. Strătilă and C. Voiculescu. VI, 594 pages. 1985.

Vol. 1133: K.C. Kiwiel, Methods of Descent for Nondifferentiable Optimization. VI, 362 pages. 1985.

Vol. 1134: G.P. Galdi, S. Rionero, Weighted Energy Methods in Fluid Dynamics and Elasticity. VII, 126 pages. 1985.

Vol. 1135: Number Theory, New York 1983–84. Seminar. Edited by D.V. Chudnovsky, G.V. Chudnovsky, H. Cohn and M.B. Nathanson. V, 283 pages. 1985.

Vol. 1136: Quantum Probability and Applications II. Proceedings, 1984. Edited by L. Accardi and W. von Waldenfels. VI, 534 pages. 1985.

Vol. 1137: Xiao G., Surfaces fibrées en courbes de genre deux. IX, 103 pages. 1985.

Vol. 1138: A. Ocneanu, Actions of Discrete Amenable Groups on von Neumann Algebras. V, 115 pages. 1985.

Vol. 1139: Differential Geometric Methods in Mathematical Physics. Proceedings, 1983. Edited by H. D. Doebner and J. D. Hennig. VI, 337 pages. 1985.

Vol. 1140: S. Donkin, Rational Representations of Algebraic Groups. VII, 254 pages. 1985.

Vol. 1141: Recursion Theory Week. Proceedings, 1984. Edited by H.-D. Ebbinghaus, G.H. Müller and G.E. Sacks. IX, 418 pages. 1985.

Vol. 1142: Orders and their Applications. Proceedings, 1984. Edited by I. Reiner and K. W. Roggenkamp. X, 306 pages. 1985.

Vol. 1143: A. Krieg, Modular Forms on Half-Spaces of Quaternions. XIII, 203 pages. 1985.

Vol. 1144: Knot Theory and Manifolds. Proceedings, 1983. Edited by D. Rolfsen. V, 163 pages. 1985.

Vol. 1145: G. Winkler, Choquet Order and Simplices. VI, 143 pag 1985.

Vol. 1146: Séminaire d'Algèbre Paul Dubreil et Marie-Paule Malliav Proceedings, 1983–1984. Edité par M.-P. Malliavin. IV, 420 pag 1985.

Vol. 1147: M. Wschebor, Surfaces Aléatoires. VII, 111 pages. 1985

Vol. 1148: Mark A. Kon, Probability Distributions in Quantum Statisti Mechanics. V, 121 pages. 1985.

Vol. 1149: Universal Algebra and Lattice Theory. Proceedings, 19 Edited by S. D. Comer. VI, 282 pages. 1985.

Vol. 1150: B. Kawohl, Rearrangements and Convexity of Level Sets PDE. V, 136 pages. 1985.

Vol 1151: Ordinary and Partial Differential Equations. Proceedin 1984. Edited by B.D. Sleeman and R.J. Jarvis. XIV, 357 pages. 198

Vol. 1152: H. Widom, Asymptotic Expansions for Pseudodifferer Operators on Bounded Domains. V, 150 pages. 1985.

Vol. 1153: Probability in Banach Spaces V. Proceedings, 1984. Edi by A. Beck, R. Dudley, M. Hahn, J. Kuelbs and M. Marcus. VI, 4 pages. 1985.

Vol. 1154: D.S. Naidu, A.K. Rao, Singular Pertubation Analysis Discrete Control Systems. IX, 195 pages. 1985.

Vol. 1155: Stability Problems for Stochastic Models. Proceedin 1984. Edited by V.V. Kalashnikov and V.M. Zolotarev. VI, 4 pages. 1985.

Vol. 1156: Global Differential Geometry and Global Analysis 19 Proceedings, 1984. Edited by D. Ferus, R.B. Gardner, S. Helgas and U. Simon. V, 339 pages. 1985.

Vol. 1157: H. Levine, Classifying Immersions into $\mathbb{R}^4$ over Stable Ma of 3-Manifolds into $\mathbb{R}^2$. V, 163 pages. 1985.

Vol. 1158: Stochastic Processes – Mathematics and Physics. Proc dings, 1984. Edited by S. Albeverio, Ph. Blanchard and L. Streit. 230 pages. 1986.

Vol. 1159: Schrödinger Operators, Como 1984. Seminar. Edited by Graffi. VIII, 272 pages. 1986.

Vol. 1160: J.-C. van der Meer, The Hamiltonian Hopf Bifurcation. 115 pages. 1985.

Vol. 1161: Harmonic Mappings and Minimal Immersions, Monteca 1984. Seminar. Edited by E. Giusti. VII, 285 pages. 1985.

Vol. 1162: S.J.L. van Eijndhoven, J. de Graaf, Trajectory Spac Generalized Functions and Unbounded Operators. IV, 272 pag 1985.

Vol. 1163: Iteration Theory and its Functional Equations. Proceedin 1984. Edited by R. Liedl, L. Reich and Gy. Targonski. VIII, 231 pag 1985.

Vol. 1164: M. Meschiari, J.H. Rawnsley, S. Salamon, Geom Seminar "Luigi Bianchi" II – 1984. Edited by E. Vesentini. VI, pages. 1985.

Vol. 1165: Seminar on Deformations. Proceedings, 1982/84. Ed by J. Ławrynowicz. IX, 331 pages. 1985.

Vol. 1166: Banach Spaces. Proceedings, 1984. Edited by N. Ka and E. Saab. VI, 199 pages. 1985.

Vol. 1167: Geometry and Topology. Proceedings, 1985. Edited b Alexander and J. Harer. VI, 292 pages. 1985.

Vol. 1168: S.S. Agaian, Hadamard Matrices and their Applications 227 pages. 1985.

Vol. 1169: W.A. Light, E.W. Cheney, Approximation Theory in Te Product Spaces. VII, 157 pages. 1985.

Vol. 1170: B.S. Thomson, Real Functions. VII, 229 pages. 1985.

Vol. 1171: Polynômes Orthogonaux et Applications. Proceedi 1984. Edité par C. Brezinski, A. Draux, A.P. Magnus, P. Maroni e Ronveaux. XXXVII, 584 pages. 1985.

Vol. 1172: Algebraic Topology, Göttingen 1984. Proceedings. Ed by L. Smith. VI, 209 pages. 1985.